Intersectional Automations

Intersectional Automations

Robotics, AI, Algorithms, and Equity

Edited by
Nathan Rambukkana

LEXINGTON BOOKS
Lanham • Boulder • New York • London

Published by Lexington Books
An imprint of The Rowman & Littlefield Publishing Group, Inc.
4501 Forbes Boulevard, Suite 200, Lanham, Maryland 20706
www.rowman.com

6 Tinworth Street, London SE11 5AL, United Kingdom

British Library Cataloguing in Publication Information Available

Library of Congress Cataloging-in-Publication Data Available

ISBN 978-1-7936-2051-4 (cloth : alk. paper)
ISBN 978-1-7936-2053-8 (paperback)
ISBN 978-1-7936-2052-1 (electronic)

♾™ The paper used in this publication meets the minimum requirements of American National Standard for Information Sciences—Permanence of Paper for Printed Library Materials, ANSI/NISO Z39.48-1992.

Contents

Acknowledgments   vii

Introduction: Intersectionality and the Machine   1

PART 1: ALGORITHMS, MACHINE LEARNING, AND INEQUITY   **15**

1   Blind Trust, Algorithmic Discrimination, and Self-Regulation in Facebook Advertisements   17
Chloé L. Nurik

2   Faking Age?: Ageing and the Algorithmic Assemblage   39
Kim Sawchuk, Scott DeJong, and Maude Gautier

3   It Was All Fun and Games: Gamewashing Automated Control   59
Sebastián Gómez

4   From Automating to Informating: Toward a Productive Model of Human/Machine Collaboration in Higher Education   77
Jordan Canzonetta

PART 2: ROBOTS AND SOCIAL JUSTICE   **97**

5   The Misogyny of Transhumanism   99
Nikila Lakshmanan

6   Are We All Too Human?: Toward an Understanding of Posthumanism and Rights   119
Julia A. Empey

7   Being Sophia: What Makes the World's First Robot Citizen?   135
        Madelaine Ley

8   Robosexuality and its Discontents   149
        Nathan Rambukkana

9   Robots as Caretakers: Understanding Long-Term
        Relationships between Humans and Carebots   169
        Jamie Foster Campbell and Kristina M. Green

PART 3: POSTHUMAN FICTIONS, FUTURES, AND BODIES   **193**

10   Im/Material Bodies: Queering Embodiment through
        Performance Art and Technology   195
        Joep Bouma

11   Estranged World: Tenets of Xenofeminism and Tropes of
        Automated Alienation in Contemporary Alien Films   213
        Christopher M. Cox

12   Simulation and Synesthesia in *Rez*: Virtual Reality and the
        Queer Erotechnics of Becoming-Machinic   237
        tobias c. van Veen

Index   261

About the Contributors   269

Acknowledgments

Like many such projects, this collection came about because of a gap in the literature. But unlike some work where the territory is all fields, the need this volume addresses was a bit different. While a lot of writing was emerging on issues such as robots, AI, algorithms, human augmentics, and power, it was scattered: often journalistic in nature, buried in broader anthologies, or centered but not intersectional in its approach. I wanted to bring together these disparate energies and to make a space for these discussions to elaborate and comingle. So, the first acknowledgment for this collection is to everyone doing work on these issues—in academia, in the industry, in policy circles, in the public broadly, and on the ground. These pieces would not be possible without everyone else who has struggled with the impact of these automated assemblages and tried to make sense of them, to bring needed critical perspectives to the discourses surrounding them.

Second, I want to thank my chapter authors, who have valiantly struggled through several years of research, peer-review, revisions, and edits to get these words to you—most of it during a global pandemic. Each chapter is a testament to the dedication of their writers who collectively wanted to foreground issues of social justice during a time of global backlash, which is not always an easy thing. Thank you as well to our excellent editors at Lexington Books and anonymous peer reviewer, who all kept working on this through lockdowns, furloughs, and working from home. The final result is stronger because of your inputs and efforts.

A third thank you goes to the many cohorts of Robotic Intimacies seminar students who, collectively, helped me navigate these emergent issues for half-a-decade. It was in the context of teaching this class that ideas about, and the need for, this collection presented themselves.

Finally, I want to thank my colleagues, mentors, friends, and family for being there during times both ridiculous and alarming. If we want to build a better future, solidarity and community are key.

This collection was in in-part supported by a Social Science and Humanities Research Council (SSHRC) Insight Development Grant, as well as an internal grant from the Faculty of Arts, Wilfrid Laurier Univeristy.

Introduction

Intersectionality and the Machine

This collection explores a range of situations where robotics, biotechnological enhancement, artificial intelligence (AI), and algorithmic culture (Striphas 2015) collide with intersectional social justice issues, such as race, class, gender, sexuality, ability, and citizenship. While selections of these subject areas and problematics might be addressed in common elsewhere, the arc this collection traces draws on a wealth of material, foci, and approaches more usually addressed in their own spheres. Part of the reason for this is to bring this book in line with the new field of Human-Machine Communication (HMC) (Guzman 2018) that draws a similar larger arc in an attempt to mobilize humanities insights from communication studies and cultural studies to more richly explore *"the creation of meaning among humans and machines"* (Guzman 2018, 1; italics in original). Another part of the reason is because this collection takes its energy from a "digital intimacy studies" (e.g., Dobson, Robards, and Carah 2018; Rambukkana and Wang 2020) that understands these diverse technological extensions of the human as both fundamentally interconnected and problematically situated with respect to issues of privilege and equity. A final reason is we find ourselves in a cultural and historical moment when these phenomena are colliding with each other, folding in on and intermingling at an increasing rate, with effects surfacing across vast topographies of life and culture.

Some call it the fourth industrial revolution (Brinded 2016; Kaplan 2015). Robots, AI, and algorithms have grown from their early uptake in some industries (such as robots in manufacturing) to an accelerating presence in other spheres ranging from customer service roles (e.g., reception, checkouts, food service, driving) to professional and creative functions previously unheard-of and un-thought-of (e.g., expert legal and medical systems, automated journalism, musical and artistic production [Kaplan 2016; Ramalho

1

2017; Hirsch 2017]). In 2016, The World Economic Forum warned that "this [would] lead to a net loss of over 5 million jobs in 15 major developed and emerging economies by 2020" (Brinded 2016), a serious challenge to ethical labor practices and a potential accelerating crisis leading some to consider alternative societal models—such as Universal Basic Income (Frase 2016), or a robot tax (Walker 2017)—to compensate.

Meanwhile, there is marked evidence that robots, AI, biotechnology, and algorithms are becoming in general and over-top of employment roles more integrated in human societies. HMC has shifted from an important yet somewhat-marginal object of study to lodge itself at the center of societal workings, economics, and visions for the future. From autonomous vehicles (Bowles 2016), to the algorithmic filtering of search results (Noble 2018) and moderation of social media content (Gillespie 2018); from online harassment and political boosterism via bots (Dewey 2016; Woolley, Shorey, and Howard 2018), to sex robots (Levy 2007; Danaher and MacArthur 2017); from ubiquitous AI assistants in our homes and smart devices (Guzman 2019), to wearable tech that tracks and shares our biometric data (Forlano 2019) and/or extends our biological capacities (Brooks 2003; Jones 2019), such technologies are rapidly mapping themselves onto almost every conceivable realm of human experience.

And yet, there is mounting evidence that the creation and programming of robots, AI, human augmentics, and algorithms, being artifacts of human culture, do not escape that context, sometimes carrying into their computational logics, platforms, and/or embodiments stereotypes, biases, exclusions, and other forms of privilege. One can think of True Companion's Roxxxy sex robots that many argue have personality types based on racist and sexist stereotypes of womanhood, for example the Barely-18 "Young Yoko" and resistant "Frigid Farah" that, as Gildea and Richardson (2017) note, seem to fetishize underage girls and sexual assault—not to mention catering to ethnic stereotypes. Or one could consider the abandoned Amazon HR algorithm. After being trained on years of resumes and hiring decisions, this machinic process used computational logic to identify traits that that were historically associated with Amazon hiring decisions, with the view of automating a stage of the hiring process, and ended up encoding a preexisting sexism from the HR data that showed that applicants with experience that included the word "Women's," or who attended all-women colleges, were often not hired (Jones 2018). One could remember poor Tay (@Tayandyou), Microsoft's short-lived teen-girl AI chatbot experiment who, after going online in March 2016, was shut down only 16 hours later after developing anti-feminist and racist personality facets through interacting with trollish users and the internet (Light 2016; Vincent 2016). Finally, one could contemplate how polities using data aggregation and predictive algorithms to manage and make decisions about social programs, resource allocation, or policing can end up targeting and profiling poor or racialized populations, with occasionally terrifying results—such as any mistake on an online application being interpreted by an automated system as "failure to cooperate" (Eubanks 2017).

We could draw on further examples of how automated technologies intersect (Crenshaw 1993) and interlock (Razack 1998) with issues of privilege and oppression, sometimes in ways that are too-little-considered and incidental, up to ways that are calculated and central to lives and communities. While this collection does not address this full scope of material, sometimes lists are important and, as Werbin (2017) notes, powerful forms. This is certainly true with respect to how they are mobilized by states as vectors of governmentality (Werbin 2017, 16), as well as by corporations who, alongside states, can now mobilize big data and machine learning in what Cheney-Lippold (2011) calls a "soft biopolitics" that assigns us "algorithmic identities." But just as top-down cultural pressures and regimes of organization can use listing and sorting as technologies to overdetermine and modulate our lives and identities, bottom-up critical work can appropriate the list form to draw an analytical circle around the issues and sites where automation and social justice intersect. Thinking through these issues together—as articulated to each other—is important. It shows us what is at stake, which elements are adjacent to others, who the multiple stakeholders are, how these issues have proliferated, and where the future might be headed.

Discussions of such issues include challenging algorithmic classism, ableism, racism, and sexism, encompassing sites such as content moderation on social media (e.g., Gillespie 2018), redlining/weblining (e.g., Eubanks 2017), business uses (e.g., Jones 2018), big data privileging (e.g., Ferguson 2017), and military practices such as Google's controversial Project Maven (e.g., Holt 2018), as well as how communities and individuals might mobilize algorithm use to address the same issues. They involve questions surrounding robotic labor and poverty, Universal Basic Income, robotic utopias and dystopias (e.g., Frase 2016; Kaplan 2016). They confront the controversies around deadly autonomous or semi-autonomous robot use by the military or non-state actors, such as work considering the Campaign to Stop Killer Robots (e.g., Anderson and Waxman 2012; Crootof 2015; Gregory 2011; Karppi, Bolen, and Granata 2016). And further controversies surrounding sex robotics, teledildonics, VR, and AI sexuality, including stereotypical, and sexist sex-robotic "personalities" and embodiments (e.g., Gildea and Richardson 2017); sex robots based on real people without consent (e.g., Gee 2017); The Campaign Against Sex Robots (e.g., Richardson 2015; Danaher, Earp, and Sandberg 2017); teledildonic/VR stream hacking and consent (e.g., Rambukkana and Gautier 2017; Belamire 2016); the interplay between robotic brothels and sex worker rights and protests (e.g., Morrish

2017; Trayner 2017; Danaher, Earp, and Sandberg 2017); bots masquerading as real people on dating sites (e.g., Light 2016; Karppi 2018); deepfakes and pornography (e.g., Maras and Alexandrou 2018), and progressive steps to address such issues or to create new sexual futures. Unavoidably they implicate the politics and ethics of the Singularity (e.g., Korb and Nicholson 2012) and the future status of robotic and AI workers with respect to labor, citizenship, and human rights—for example, work on Hansen Robotics' Sophia as Saudi citizen (e.g., Weller 2017), robotic servitude (e.g., Green 2016), as well as the rights of humans interacting with AI (e.g., Shepherd 2019). And crucially, they have to address the assumptions, representations, and discourse surrounding dis/ability and human augmentics, including "supercrip" and "cyborg" discourses, as well as the potential tensions between feminist technoscience (e.g., Haraway 1990) and critical disability studies (e.g., Allan 2016; Cascais 2013). Finally, they would need to take into account how these and adjacent issues are depicted in popular or fringe fictions that contain robotic or AI characters, such as *Humans* (Lundström 2015), *Neuromancer* (Gibson 1984), *Extant* (Fisher 2014), *Westworld* (Joy and Nolan 2016), *Her* (Jonze 2013), *Blade Runner* (Scott 1982), *Ex Machina* (Garland 2014), *Ghost in the Shell* (Oshii 1995), *Altered Carbon* (Kalogridis 2018), *Black Mirror* (Brooker 2011), *Neon Genesis Evangelion* (Anno 1995), and *Questionable Content* (Jacques 2003). How these issues are represented, reified, or refigured in art and fiction speak not only to how we read and interpret the present, but how we are hoping to—or fearing we might—inscribe the future.

This collection draws such an analytical circle around these interconnected and adjacent issues, lending a critical eye to a modest selection of problematics that, collectively, explore what is at stake due to the automation of aspects of culture. How do equity issues intersect with these fields? Are the pronouncements always already dire, or are there also lines of flight toward more equitable futures in which agentic artifacts and extensions can play an active part?

BREAKDOWN OF SECTIONS

The fifteen authors assembled here explore many of the above issues in three interrelated sections. The first looks at fundamental but sometimes ephemeral issues surrounding Algorithms, Machine Learning, and Inequity; the second takes on the more embodied problematics of Robots and Social Justice; and the final section, Posthuman Fictions, Futures, and Bodies, looks at how some of these themes are being explored through artistic media.

Algorithms, Machine Learning, and Inequity

From what pops up in our Facebook feeds or Netflix recommends, to the kinds of products, services, and activities that are advertised to us, to how machine learning enters our lives as tools for work or leisure, "algorithmic culture" (Striphas 2015) has made a significant intervention into our daily lives. The four chapters in this section touch on very different domains of experience from social media, to bio-informatic wearables, to EdTech, yet they converge on sketching out a narrative in which nonhuman actants are taking their places as agents in and among human (sociotechnical) networks (Latour 2005).

In chapter 1, Chloé Nurik documents how Facebook, while holding to the letter of U.S. regulations prohibiting digital redlining for certain types of advertising (e.g., related to housing, credit, or employment), allows advertisers to privilege proxy variables (such as zip codes) that have materially similar effects. Furthermore, as she establishes through a set of interviews with relevant stakeholders, as the process of algorithmic selection is both blackboxed and largely self-regulated, the variables chosen by advertisers are further mixed and nuanced by "algorithmic identities" (Cheney-Lippold 2011) and "algorithmic inferences" (Nurik, this book), and in ways that may indirectly discriminate and disenfranchise specific groups—such as excluding disabled people by allowing advertisers to indicate they do not want to target those the algorithm identifies as having an "interest in disability rights." She argues for a stronger regulatory regime to bring the practice of algorithm-assisted advertising back in line with the spirit of laws already in place.

Exploring a more focused subset of the same issues, in chapter 2 Kim Sawchuk, Scott DeJong, and Maude Gauthier take a bottom-up approach of reverse engineering profiles for Facebook users who present as different ages and comparing and contrasting how they are addressed by algorithms to how the researchers themselves are targeted by suggested posts and ads on their personal accounts. Their experiment establishes not only how Facebook algorithms make certain assumptions about their interests and concerns depending on how the algorithm reads—as well as assumptions about—age and gender, but also that this type of research is complicated by machine learning's keen ability to spot what might be fake or duplicate profiles such as the ones they created, and to either read-around them (as ascertained by targeting interests to the experimental profiles that would be more apt to fit personal profiles accessed from the same devices) and even the disabling of one of the accounts. This last methodological insight is key with respect to being strategic about how we approach researching algorithms and AI, as this chapter shows that via the meta-network of analytics, machine learning is neither merely a neutral technic nor incapable of resisting overt scrutiny. Put another way, analyzing multifaceted data is very much its wheelhouse.

Moving from social media to gamification and wearables, in chapter 3 Sebastián Gómez uses critical posthumanism to unpack the ways that devices and apps that work to shape human behavior could be seen as process of "gamewashing control," in that inviting technological devices to quantify our biodata, record it, sort it, and set out incremental goals for us can take part in a process of "humanist normalization" (Gómez, this book). What agentic ground are we ceding when we allow ourselves to buy into the crafted ludic goals of achieving badges or passing levels. His chapter is a caution that even gamifying what might seem like neutral personal goals "in the name of fun" (such as using a Fitbit tracker to motivate you to be more active) might have unexamined normativities articulated to it, such as organizations mandating employees wear them if they want to reduce their insurance premiums (Booton 2015).

Finally, in chapter 4, Jordan Canzonetta assesses machine learning–supported educational technology with a view to parsing out both the negative and positive use-cases and designs for such technologies. Pushing back slightly on the overdetermined "Fourth Industrial Revolution" narrative that the machines are going to take our jobs (e.g., Brinded 2016), Canzonetta looks at how EdTech, such as Turnitin and Eli Review, while at risk for becoming cold, automating, alienating, and anti-social technologies, could be designed and used rather to "informate" lower-level pedagogical tasks in ways that, instead of replacing teachers, can free up their time and energy for more important forms of student engagement. Interviewing both the technological objects themselves as well as teachers who have used them, she tracks a possible path forward in which humans and machine learning can be true partners in education, as opposed to rivals.

Robots and Social Justice

The five chapters in this section dig into the techno-futurist world of robots, human augmentics, and even eugenics to uncover how issues of equity and justice play out in this broad field. From the theoretical and movement tensions between (and within) posthumanism and transhumanism; to negotiating the discursive, ontological, and political prominence Hanson Robotics' Sophia; to the controversies surrounding sex robotics as a practical field that is now upon us; to important considerations for care robotics, these chapters sample diverse issues, while at the same time converging around questions of what "the good" is in relation to the role of robotics in society, as well as considering how far we are from those ideals.

In chapter 5, Nikila Lakshmanan does a deep dive into transhumanism. In deconstructing the ideological movement of transhumanism and radically situating it with respect to its uncomfortable cousin posthumanism and its (surprisingly close) parent discourse humanism, she parses a more problematic "libertarian transhumanism" from a posthuman-inspired "democratic transhumanism." Lakshmanan argues that the latter has a potential to push against the boundaries of what we consider human (e.g., engaging with robotics, human augmentics, and extensions) without the pitfalls of reifying discourses such as those of eugenics, misogyny, and the far right, which all have uncomfortably high levels of expression in libertarian transhumanism.

In chapter 6, Julia A. Empey also engages with posthumanism in the context of rights and in relation to star robot spokes-"person," and nominal Saudi Arabian citizen, Sophia. She notes, "By rethinking the role of the human, one needs to explore what happens to the 'human' part of human rights" (Empey, this book). Like the previous chapter's parsing of transhumanism, this chapter contrasts what we might be tempted to call a normative posthumanism (often infused with libertarian transhumanist ideology and shorthanded in this chapter as "the posthuman") with a "critical posthumanism" which, like with the democratic transhumanism discussed by Lakshmanan, attempts not to create a transcendent human supremacy but rather rethink the role of the human with respect to its contexts such as the world, nature, and human rights.

Madelaine Ley's chapter 7 also interrogates Sophia and its[1] role in media, politics, and popular culture broadly. She sees Sophia as a concretized embodiment—or manifestation—of particular choices, ideologies, and desires. Sophia is an ideological tool used to sell the idea of a particular kind of technosocial future—or even convince us that is it already here, with Sophia as its spokes-"person." Ley argues this is a subjunctive move, a *trompe-l'œil* synecdochally linked to presenting it as "her," and ironically foregrounded by elevating that "her" to being a subject with rights in a Saudi Arabia that both wants to portray itself as progressive and technologically cutting edge, and where its actual female subjects often struggle with regressive laws and policy. Sophia's non-threatening gendered and raced presentation makes it (*qua* "her"), problematically, the perfect figure to "speak for the future"—no matter how many times "she" casually drops the notion of "human zoos" into conversation.

In chapter 8, Nathan Rambukkana looks at the controversies surrounding human–robot sexuality—or, as the article frames it ripping a term from science fiction, "robosexuality." Cutting across the debate between robosexuality advocates like David Levy and robosexuality skeptics like Kathleen Richardson, Rambukkana charts a middle path that acknowledges that both sides have a mixture of important points as well as dangerously oversimplified rhetorical stances. In the end, he argues that it is not straightforward to map robosexuality onto social justice concerns, in-part because, crucially, sex robotics as currently manifest in industry are just a small subset both of what a more queer and feminist-informed egalitarian sex robotics might look like, as well as of what robo-sexuality might mean in the future when we are talking prospectively about the sexual choices of robotic beings and not simply fully embodied sex toys.

Finally, in chapter 9, Jamie Foster Campbell and Kristina M. Green review how social robots might become a more extensive part of human societies, and in particular how carebots "will and can be used for long-term caregiving" (Foster Campbell and Green, this book). Rather than argue the question of if carebots should be put in those roles, Foster Campbell and Green speak to their current and likely future deployment to meet the needs of a rapidly aging population, and urge that an attention to interpersonal communication scholarship and, in particular, work on reciprocal trust and self-disclosure, will be some of the keys to both successful and just HMC relationships in healthcare settings, where they will also have to negotiate thorny ethical decisions such as those around end-of-life decisions that, as Karppi, Bolen, and Granata (2016) note, place carebots on a continuum with autonomous cars and weapons systems, in that they might be called upon to decide if a person lives or dies.

Posthuman Fictions, Futures, and Bodies

The three chapters in the final section of the collection explore cultural products from performance art and music, to Hollywood films, to video games to track the ways that the realm of art and creation interacts with the ideologies and visions of automation. While fictional works treating robots and AI are also addressed in other sections (notably in chapters 6 and 8), here the authors look broader to incorporate queer and trans embodiment, the movement of metaphor and allegory, and queer erotechnics.

In chapter 10, Joep Bouma curates a selection of works from queer and trans artists that, together, explores their use of digital technologies to create experimental embodiments that materialize a queer posthumanism. At the intersection of gender performativity and performance art, Bouma locates diverse technologically mediated encounters with, or extensions of, the body through technological means. With multiple works from six artists which include a music album, music videos, a musical dildo, performance art pieces, and a video piece, this chapter foregrounds the potential of a queer cyborg body that pushes notions of becoming into a continual state of open-ended technological hybridity in which "realness" is multiple, contingent, non-fixed, and fluid.

Christopher Cox looks at science fiction takes on alien others as allegories of our attitudes toward automated otherness in chapter 11. In unpacking the implications for xenofeminism in light of how the narratives of three alien encounter films play out, Cox's imminent critique of this strand of open-ended techno-feminism is an important reminder that ideological discourses sometimes have hard limits and progressive narrative might not track perfectly onto practice—at least not comfortably. For example, how does xenofeminism's embrace of alienation as emancipatory sit with narratives of literal alien encounters which defamiliarize normative time while at the same time reifying gender essentialism? If science and speculative fiction generate potentially aspirational, potentially cautionary narratives, what do we do when ideological positionalities we might project on them end up mapping onto both?

Finally, in chapter 12, tobias c. van Veen unpacks the "queer erotechnics" of the synesthetic *Rez* videogame series, in which your character, a hacker, moves through a 1990s-style representation of virtual reality, breaking into a mainframe through assembling rave style music (similar to *Guitar Hero*), but in a way that the music becomes continuous with not only the visuals (colors and patterns) but also haptics through an overtly sexually charged "Trance Vibrator" peripheral. Becoming the avatar in a simulation of both VR and of rave culture queers the experience of gaming due to its gender fluid embodiments and furthers the synesthetic ability to "feel the music" as part of a hybrid human/nonhuman desiring-machine: an erotechnical cyborg.

Together, these chapters form a tour of some of the cultural sites where social justice and automation intersect. While only a fragment of code in the mainframe, we see this collection as a conversation starter, as well as an invitation to continue to research and have important discussions about these issues, and always with one eye to questions of equity, moving forward.

NOTE

1. Ley deliberately does not gender Sophia as female to emphasize how Sophia does not in her view legitimately embody any human gender. See also note 1, chapter 7.

REFERENCES

Allan, Kathryn. 2016. "Categories of Disability in Science Fiction." *Kathryn Allan's Blog* (blog), January 27, 2016. http://www.academiceditingcanada.ca/blog/item /317-disability-in-sf-article.

Anderson, Kenneth, and Matthew Waxman. 2012. "Law and Ethics of Robot Soldiers." *Policy Review*, December 2012 and January 2013. Columbia Public Law Research Paper No. 12–313.American University, WCL Research Paper No. 2012–32.

Anno, Hideaki, creator. 1995. *Neon Genesis Evangelion*. Gainax. Aired on Animax.

Belamire, Jordan. 2016. "My First Virtual Reality Groping." *Mic*, October 21, 2016. https://mic.com/articles/157415/my-first-virtual-reality-groping-sexual-assault-in -vr-harassment-in-tech-jordan-belamire - .FubnFVP5F.

Booton, Jennifer. 2015. "You May Be Forced to Wear a Health Tracker at Work." *MarketWatch*, March 12, 2015. https://www.marketwatch.com/story/you-might-be -wearing-a-health-tracker-at-work-one-day-2015-03-11.

Bowles, Nellie. 2016. "Google Self-driving Car Collides with Bus in California, Accident Report Says." *Guardian*, March 1, 2016. https://www.theguardian.com/ technology/2016/feb/29/google-self-driving-car-accident-california.

Brinded, Lianna. 2016. "WEF: Robots, Automation, and AI will Replace 5 Million Human Jobs by 2020." *Business Insider*, January 19, 2016. https://www.business insider.com.au/wef-davos-report-on-robots-replacing-human-jobs-2016-1.

Brooker, Charlie, creator. *Black Mirror*. 2011. Zeppotron and Channel 4 Television Corporation. Netflix.

Brooks, Rodney. 2003. *Flesh and Machines: How Robots Will Change Us*. New York: Vintage.

Cascais, António. F. 2013. "The Metamorphic Body in Science Fiction: From Prosthetic Correction to Utopian Enhancement." In *Disability in Science Fiction*, edited by Katharyn Allan, 61–72. New York: Palgrave Macmillan.

Cheney-Lippold, John. 2011. "A New Algorithmic Identity: Soft Biopolitics and the Modulation of Control." *Theory, Culture & Society* 28, no. 6: 164–181. https://doi .org/10.1177/0263276411424420.

Crenshaw, Kimberlé. 1996. "Mapping the Margins." *Stanford Law Review* 43, no. 1231: 1241–1299.

Crootof, Rebecca. 2015. "War, Responsibility, and Killer Robots." *North Carolina Journal of International Law and Commercial Regulation* 40, no. 4: 909–932.

Danaher, John, Brian Earp, and Anders Sandberg. 2017. "Should We Campaign Against Sex Robots?" In *Robot Sex: Social and Ethical Implications*, edited by John Danaher and Neil MacArthur, 47–72. Cambridge, MA: MIT Press.

Danaher, John, and Neil MacArthur, eds. 2017. *Robot Sex: Social and Ethical Implications*. Cambridge, MA: MIT Press.

Dewey, C. 2016. "One in Four Debate Tweets Comes from a Bot. Here's How to Spot Them." *Washington Post*, October 19, 2016. https://www.washingtonpost .com/news/the-intersect/wp/2016/10/19/one-in-four-debate-tweets-comes-from-a -bot-heres-how-to-spot-them/?utm_term=.a757c59bc072.

Dobson, Amy Shields, Brady Robards, and Nicholas Carah, eds. 2018. *Digital Intimate Publics and Social Media*. London: Palgrave Macmillan.

Eubanks, Virginia. 2017. *Automating Inequality: How High-Tech Tools Profile, Police, and Punish the Poor*. New York: St. Martin's Press.

Ferguson, Andrew G. 2017. *The Rise of Big Data Policing: Surveillance, Race, and the Future of Law Enforcement*. New York: New York University Press.

Fisher, Mickey, creator. *Extant*. 2014. 21 Plates. CBS.

Forlano, Laura. 2019. "Posthuman Futures: Connecting/Disconnecting the Networked (Medical) Self." In *A Networked Self and Human Augmentics, Artificial Intelligence, Sentience*, edited by Zizi Papacharissi, 39–50. New York: Routledge.

Frase, Peter. 2016. *Four Futures: Life after Capitalism*. London: Verso.

Garland, Alex, dir. 2014. *Ex Machina*. USA: Universal Pictures.

Gee, Tabi J. 2017. "Why Female Sex Robots are more Dangerous than you Think." *Telegraph*, April 28, 2017. https://www.telegraph.co.uk/women/life/female-robots -why-this-scarlett-johansson-bot-is-more-dangerous/.

Gibson, William, 1984. *Neuromancer*. New York: Ace Science Fiction Books.

Gildea, Florence, and Kathleen Richardson. 2017. "Sex Robots: Why We Should be Concerned." *Campaign Against Sex Robots*, May 12, 2017. https://campaignagai nstsexrobots.org/2017/05/12/sex-robots-why-we-should-be-concerned-by-florence -gildea-and-kathleen-richardson/.

Green, Shelleen M. 2016. "Bina48: Gender, Race, and Queer Artificial Life." *Ada: A Journal of Gender, New Media & Technology* 9. https://adanewmedia.org/2016 /05/issue9-greene/.

Gregory, Derek. 2011. "From a View to a Kill Drones and Late Modern War." *Theory, Culture & Society* 28, no. 7–8: 188–215. https://doi.org/10.1177 /0263276411423027.

Guzman, Andrea. L. Ed. 2018. *Human-Machine Communication: Rethinking Communication, Technology, and Ourselves*. New York: Peter Lant.

———. 2019. "Beyond Extraordinary: Theorizing Artificial Intelligence and the Self in Daily Life." In *A Networked Self and Human Augmentics, Artificial Intelligence, Sentience*, edited by Zizi Papacharissi, 84–96. New York: Routledge.

Haraway, Donna. 1990. "A Manifesto for Cyborgs: Science, Technology, and Socialist Feminism in the 1980s." In *Feminism/postmodernism*, edited by Linda J. Nicholson, 190–233. New York: Routledge.

Hirsch, Peter B. 2017. "The Robot in the Window Seat." *Journal of Business Strategy* 38, no. 4, 47–51. https://doi.org/10.1108/JBS-04-2017-0050.

Holt, Kris. 2018. "Google Employees Reportedly Quit over Military Drone AI Project." *Engadget*, May 14, 2018. https://www.engadget.com/2018/05/14/google -project-maven-employee-protest/.

Jacques, Jeph. 2003. *Questionable Content* (webcomic). https://questionablecontent .net.

Jones, Rhett. 2018. "Amazon's Secret AI Hiring Tool Reportedly 'Penalized' Resumes with the Eord 'Women's'." *Gizmodo*, October 10, 2018. https://gizmodo .com/amazons-secret-ai-hiring-tool-reportedly-penalized-resu-1829649346.

Jones, Steve. 2019. "Untitled, no. 1 (Human Augmentics)." In *A Networked Self and Human Augmentics, Artificial Intelligence, Sentience*, edited by Zizi Papacharissi, 201–205. New York: Routledge.

Jonze, Spike, dir. 2013. *Her*. USA: Warner Brothers.

Joy, Lisa, and Jonathan Nolan, creators. 2016. *Westworld*. Bad Robot. HBO.

Kalogridis, Laeta, creator. *Altered Carbon*. 2018. Mythology Entertainment and Skydance Television. Netflix.

Kaplan, Jerry. 2015. "Robots are Coming for your Job: We must Fix Income Inequality, Volatile Job Markets Now—Or Face Sustained Turmoil." *Salon*, August 23, 2015. https://www.salon.com/2015/08/23/robots_are_coming_for_your_job_we_must_fix_income_inequality_volatile_job_markets_now_or_face_sustained_turmoil/.

Karppi, Tero. 2018. "'How Angels are Made': Ashley Madison and the Social Bot Affair." In *A Networked Self and Love*, edited by Zizi Papacharissi, 173–188. New York: Routledge.

Karppi, Tero, Marc Bölen, and Yvette Granata. 2016. "Killer Robots as Cultural Techniques." *International Journal of Cultural Studies*, October 16. https://doi.org/10.1177%2F1367877916671425.

Korb, Kevin B. and Anne E. Nicholson. 2012. "Ethics of the Singularity." *Issues*, March, 2012. http://www.issuesmagazine.com.au/article/issue-march-2012/ethics-singularity.html.

Latour, Bruno. 2005. *Reassembling the Social: An Introduction to Actor-Network-Theory*. Oxford: Oxford University Press.

Levy, David, 2007. *Love + Sex with Robots: The Evolution of Human—Robot Relationships*. New York: Harper.

Light, Ben. 2016. "The Rise of Speculative Robots: Hooking up with the Bots of Ashley Madison." *First Monday: Peer-Reviewed Journal on the Internet* 6, no. 6. http://journals.uic.edu/ojs/index.php/fm/article/view/6426.

Lundström, Lars, creator. 2015. *Humans*. Kudos. Channel 4.

Maras, Marie-Helen, Alex Alexandrou. 2018. "Determining Authenticity of Video Evidence in the Age of Artificial Intelligence and in the Wake of Deepfake Videos." *International Journal of Evidence & Proof*, October 28. https://doi.org/10.1177/1365712718807226.

Morrish, Lydia. 2017. "A Sex Doll Brothel Is Set To Open In The UK." *Konbini*, April 28, 2017. http://www.konbini.com/en/lifestyle/sex-doll-brothel-uk/.

Noble, Safiya Umoja. 2018. *Algorithms of Oppression: How Search Engines Reinforce Racism*. New York: NYU Press.

Oshii, Mamoru, dir. 1995. *Ghost in the Shell*. Bunkyō, Tokyo: Kodansha.

Ramalho, Ana. 2017. "Will Robots Rule the (Artistic) World?: A Proposed Model for the Legal Status of Creations by Artificial Intelligence Systems." *Journal of Internet Law* 21, no. 1: 11–25.

Rambukkana, Nathan, and Keer Wang. 2020. "Digital Intimacies." In *Oxford Bibliographies Online*. https://www.oxfordbibliographies.com/view/document/obo-9780199756841/obo-9780199756841-0250.xml.

Rambukkana, Nathan, and Maude Gautier. 2017. "L'Adultère à l'Ère Numérique: Une Discussion sur la Non/Monogamie et le Développement des Technologies Numériques à Partir du cas Ashley Madison." *Genre, Séxuality et Société* 17. https://journals.openedition.org/gss/3981.

Razack, Sherene H. 1998. *Looking White People in the Eye: Gender, Race, and Culture in Courtrooms and Classrooms*. Toronto: University of Toronto Press.

Richardson, Kathleen. 2015. "Welcome to the Campaign Against Sex Robots." *Campaign Against Sex Robots*, September 15, 2015. https://campaignagainstsexrobots.org/2015/09/15/welcome-to-the-campaign-against-sex-robots/.

Scott, Ridley, dir. 1982. *Blade Runner*. USA: Warner Brothers.

Shepherd, Tamara. 2019. "AI, the Persona, and Rights." In *A Networked Self and Human Augmentics, Artificial Intelligence, Sentience*, edited by Zizi Papacharissi, 187–200. New York: Routledge.

Striphas, Ted. 2015. "Algorithmic Culture." *European Journal of Cultural Studies* 18, no. 4–5: 395–412. https://doi.org/10.1177/1367549415577392.

Trayner, David. 2017. "First Sex Doll Brothel in Europe Shut Down One Month after Opening before Police Raid." *Daily Star*, March 17, 2017. https://www.dailystar.co.uk/news/latest-news/597431/lumidolls-europe-first-sex-robot-love-doll-brothel-barelona-spain-closed-shut-down-police.

Vincent, James. 2016. "Twitter Taught Microsoft's AI Chatbot to be a Racist Asshole in Less than a Day." *The Verge*, March 4, 2016. https://www.theverge.com/2016/3/24/11297050/tay-microsoft-chatbot-racist.

Walker, Jon. 2017. "Robot Tax—A Summary of Arguments 'For' and 'Against.'" *Tech Emergence*, October 24, 2017. https://emerj.com/ai-sector-overviews/robot-tax-summary-arguments/.

Weller, Chris. 2017. "A Robot that Once Said It Would 'Destroy Humans' just Became the First Robot Citizen." *Business Insider*, October 26, 2017. https://www.businessinsider.com/sophia-robot-citizenship-in-saudi-arabia-the-first-of-its-kind-2017-10.

Werbin, Kenneth C. 2017. *The List Serves: Population Control and Power*. Amsterdam: Institute of Network Cultures.

Woolley, Samuel, Samantha Shorey, and Philip Howard. 2018. "The Bot Proxy: Designing Automated Self Expression." In *A Networked Self and Platforms, Stories, Connections*, edited by Zizi Papacharissi, 59–76. New York: Routledge.

Part 1

ALGORITHMS, MACHINE LEARNING, AND INEQUITY

Chapter 1

Blind Trust, Algorithmic Discrimination, and Self-Regulation in Facebook Advertisements

Chloé L. Nurik

Targeting consumers is a standard advertising practice that is largely perceived as efficient and effective (Turow 2011). While this practice predates the Internet and the growth of algorithms (Baker 1994; Spurgeon 2007), the potential for unlawful discrimination on weakly regulated social media platforms is tremendous. In particular, Facebook—with its wealth of user data and sophisticated targeting tools—has enabled advertisers to identify potential consumers with pinpoint precision while simultaneously blocking other people from viewing messages (Rieke and Bogen 2018). Although it is illegal in the United States (U.S.) to limit access to housing, credit, or employment advertisements based on protected classes (i.e., color, disability, familial status, national origin, race, religion, and sex), offenses that occur online have been shielded by Facebook's complex, opaque, and profit-driven operations (Allen 2019).

In recent years, the exclusionary algorithmic hypertargeting of commercial messages on Facebook has come to light. In 2016, *ProPublica* reported that Facebook permitted housing advertisers to exclude users by race (Angwin and Parris 2016). Over the next three years, Facebook promised to reform no less than seven times; however, racial exclusions from housing ads as well as age- and gender-based discrimination in employment ads continued on the platform (Angwin, Scheiber, and Tobin 2017; Scheiber 2018). While Facebook eventually reached a settlement with advocacy groups and faced formal charges from the U.S. Department of Housing and Urban Development (HUD) in 2019, the issue of algorithmic advertising discrimination on the platform remains cause for concern, especially since advertising constitutes the site's primary source of profit (Fuchs 2012). Advertising discrimination is thus an outgrowth of Facebook's business model that treats users as commodities, a structural underpinning that complicates the platform's repeated promises to curb discriminatory advertising. Given Facebook's unparalleled global influence as well as the dramatic consequences of exclusionary targeting, it is critical to examine advertising discrimination and its algorithmic elements to pave the way for more equitable policies and regulations that will force Facebook to look beyond its bottom line and to instead prioritize protection of its users.

In the contemporary "algorithmic society" (Balkin 2016, 1232), algorithms have become embedded in everyday life, from finance (Chen 2012) to the criminal justice system (Murphy 2007). While early work on algorithms highlighted their potential for circumventing human biases, recent literature has analyzed how algorithms replicate the discriminatory patterns they were intended to transcend, building upon stereotyped notions of class, race, and gender (DeVito 2017; Eubanks 2017; Noble 2018; Martin 2018). Furthermore, researchers have lamented the existence of technological "black boxes" (Pasquale 2015, 3), in which operating procedures are cloaked in secrecy. Studies in this area have also examined the way algorithms can trigger negative societal consequences, such as filter bubbles (Bozdag and van den Hoven 2015), echo chambers (Tufekci 2015), discrimination (Noble 2018), or weblining and exclusion from the digital marketplace (Danna and Gandy 2002; Eubanks 2017). Several recent studies have specifically addressed algorithms used in social media advertising (Datta et al. 2018; Speicher et al. 2018). However, the existing literature mostly pertains to biases in algorithmic design or application rather than regulatory issues that both create and perpetuate bias and inequity, an increasingly important area of concern.

INTERVIEWS AND ARGUMENTS: STUDYING ADVERTISING DISCRIMINATION

To fill the aforementioned gap, this chapter provides a detailed case study of algorithmic advertising discrimination on Facebook. In particular, I outline the technical mechanisms, social impacts, and regulatory underpinnings of such discrimination. Direct observation of advertising discrimination is not possible since users do not know what ads they are excluded from seeing, and Facebook is purposefully opaque about its advertising practices to avoid public scrutiny and legal sanctions (Rieke and Bogen 2018; Kim 2020).[1] Consequently, I conducted semi-structured elite interviews (Beyers et al. 2014) and informant interviews with marketers, lawyers, researchers, and academics between September and November 2018 to learn more about the mechanics, effects, and regulation of advertising discrimination on Facebook.

Table 1.1 Participants

Interviewee	Occupation	Interview Format	Date
Chase Morris	Marketer	Phone call	September 26, 2018
Gwen Stewart	Lawyer (civil rights)	Phone call	September 28, 2018
Matthew Walker	Lawyer (housing)	Phone call	October 1, 2018
Andrew Reyes	Researcher (technology and social justice)	Phone call	October 5, 2018
Raymond Saunders	Law professor (housing)	Phone call	October 10, 2018
Orlando Simon	Law professor (information and communications)	Phone call	October 15, 2018
Timothy James	Advertising professor	Face-to-face	October 16, 2018
Ben Rollins	Marketing professor	Face-to-face	October 19, 2018
Alex Mills	Computer science professor (privacy and security)	Phone call	October 23, 2018
Mason Taylor	Researcher (privacy and security)	Phone call	October 23, 2018
Caleb Thomas	Law professor (privacy)	Phone call	November 1, 2018

At the time of this study, two interviewees were actively involved in legal proceedings against Facebook regarding the platform's facilitation of discriminatory advertising. Table 1.1 lists the pseudonym and profession of interviewees as well as the interview format and date.

To supplement the interviews, I examined an extensive corpus of documents, including Facebook's advertising policies, news articles about advertising discrimination on social media, and case material related to current lawsuits against Facebook.

Drawing from these sources, I make three main arguments in this chapter: First, countering simplistic claims of causality, advertising discrimination on Facebook is the outgrowth of several factors, including problematic algorithms, problematic tendencies baked into capitalist media systems, and problematic features of self-regulation. Second, while academic discussions about algorithms have largely sidestepped the question of regulation, it is essential to consider systems of regulation when studying and reforming algorithms. We cannot effectively grapple with the reach and impacts of algorithms unless we foreground issues of regulation that give rise to prejudicial ones and enable abusive behavior to continue. Third, in the current moment, it is untenable to suggest that no regulation or even self-regulation of algorithms is all that is required (Reed 2018). As will be demonstrated in this case study of Facebook's advertising system, regimes of self-regulation have paved the way for discrimination, harming marginalized users and exacerbating offline inequities (Nurik 2019).

HOW SELF-REGULATION ENABLES SOCIAL
MEDIA SITES TO EXPLOIT LEGAL LOOPHOLES

Social media platforms such as Facebook are regulated in a distinct manner from other media in the United States as the former are chiefly self-regulated, meaning industry creates and enforces its own norms and standards with limited government intervention (Nurik 2018). The primary piece of U.S. legislation shaping social media governance is the Communications Decency Act (CDA) of 1996. § 230 of the CDA limits liability for interactive computer services (such as platforms) and enables these entities to develop their own regulatory approaches. Specifically, § 230 stipulates that platforms are not liable for either hosting *or* removing user-generated content.

This legal protection has triggered unintended consequences, including unequal enforcement and discriminatory patterns in platform governance (Nurik 2019). Most significantly, the broad immunity conferred in § 230 directly conflicts with platforms' legal obligations to adhere to civil rights laws in the United States (Collins 2008). A number of civil rights laws (including the Civil Rights Act of 1964 (§ 2000e-3), the Age Discrimination in Employment Act of 1967 (§ 623), the Civil Rights Act of 1968 (§ 1901.203), and the Equal Credit Opportunity Act of 1974 (§ 1691)) make it illegal for employers, creditors, and landlords (as well as the media they advertise in) to print or cause to be printed advertisements that discriminate on the basis of race, color, religion, sex, national origin, age, disability, and familial/marital status.

Although discriminatory advertisements have been outlawed for decades, digital advertising has brought forth numerous complications and opportunities for evading legal restrictions. The static provisions in civil rights laws have been rendered ambiguous and uncertain as the lines between distributor and publisher, advertising and content, and discrimination and normal algorithmic functioning have become blurred (Sweeney 2013; Ferrer-Conill et al. 2020). The shifting nature of these terms and categories makes it difficult to establish definitive legal standards. The unresolved nature of these questions is particularly problematic given the seeming contradiction between civil rights laws and § 230 of the CDA (Collins 2008). How can platforms be immunized for user-generated content *and yet* be prohibited from publishing discriminatory advertisements? The conflict between these laws has perpetuated advertising discrimination on Facebook as the platform takes advantage of uncertain and evolving guidelines.

DISCRIMINATING ADVERTISERS, DISCRIMINATING ALGORITHMS

Individuals do not pay to create or use a Facebook account. Consequently, users themselves become the commodity as their time and attention are commodified (Smythe 1977) and sold to advertisers for a hefty profit (Baker 1994). While making money *from* advertisers, Facebook makes money *off* users through an exploitative and asymmetric relationship (Fuchs 2012). Thus, Facebook's advertising system is structured to benefit both advertisers and the platform at the expense of users whose information and attention are utilized to generate profit.

Given wide latitude because of its self-regulatory governance, Facebook created intricate targeting options that allow it to sift through users, offering advertisers the unparalleled ability to "reach the right people" (Facebook, n.d. [d]). These sophisticated tools are predicated upon segmentation and segregation: "At the heart of Facebook's powerful marketing platforms are tools for advertisers to exclude or include Facebook users who will view an advertisement [. . .] based on various protected classes" ("First amended complaint" 2017, 5). Discrimination is therefore woven throughout Facebook's advertising system, partially constituting its guiding logic and means of profitability. The platform's involvement in the advertising process exists on a continuum from providing the infrastructure advertisers use to scouting out potential audiences for advertisers (Rieke and Bogen 2018). The determination of which users see ads is the result of ad targeting procedures and ad delivery processes. Of note, algorithms are operative in the construction of audience attributes (for core audiences), the creation of lookalike audiences, and the processes of ad delivery.

A key method to "slice and dice audiences and groups" (Caleb, interview, November 1, 2018) is the use of core audiences, wherein an advertiser selects audiences according to demographics, location, interests, and behaviors (Facebook, n.d. [c]). Targeting is based on both user-provided information (when constructing a profile) such as gender, and inferred categories such as liberal political orientation (Rieke and Bogen 2018). Inferred categories are produced by Facebook's machine learning algorithms, which analyze an extensive amount of user data across the site (and across other websites) to create probable user characteristics. These algorithmic inferences can lead to "indirect discrimination" (Speicher et al. 2018, 8) in advertising targeting. For example, Facebook previously offered (and now claims to prohibit) the following targeting categories: "young and hip" ("First amended class and collective action complaint" 2018, 22), "millennials" ("First amended class and collective action complaint" 2018, 22), "interest in disability rights" (Tobin 2018, para. 11), "soccer mom" ("Charge of discrimination" 2018, 13), and "single dads" ("Charge of discrimination" 2018, 13). These options, programmed by sophisticated algorithms, give advertisers the ability to discriminate against various protected classes by using proxy variables. Of course, advertisers can also directly discriminate by simply using demographic information to exclude protected classes without resorting to proxy variables.

There are two additional ways to target audiences. Facebook allows advertisers to create custom audiences, a specific list of people the advertiser has already engaged with and/or has specific information about from sources such as in-store interactions, meaning advertisers can target individual users (Facebook, n.d. [b]). Custom audiences can be enhanced through the use of lookalike audiences, individuals Facebook's algorithms determine to be similar to the initial custom audience (Facebook, n.d. [b]). While Facebook does not provide information about which factors its algorithms consider when constructing a lookalike audience, it is suspected that proxies for protected classes are used. In fact, through a series of experiments, Speicher et al. (2018) found that the racial or political affiliation of a custom audience (that the advertiser produced) can be replicated in resulting lookalike audiences created by algorithms, underscoring the role of both advertisers and the platform in creating "a discriminating monstrosity" (Caleb, interview, November 1, 2018).

While Facebook's algorithmically derived categories and advertisers' explicit choices can lead to discrimination in targeting, discrimination can also occur in the ad delivery phase due to algorithmic predictions. Facebook can "dispatch[] ads with surgical precision" (Rieke and Bogen 2018, 8) based on the results of an instantaneous auction that occurs behind the scenes.[2] Whenever a user is online and there is an empty content slot, an auction occurs (billions of auctions occur each day) in which advertisements compete to be shown (Facebook, n.d. [a]). The winner of the auction is determined by the dollar amount of the bid and Facebook's algorithmic estimate of how interested a person will be in the ad based on its overall quality and relevance to the user. After an advertising campaign starts running, Facebook collects data about who is interacting with the ad and refines its predictions of the right person to show the ad to. For example, even if an advertiser did not specify gender-based targeting (a discriminatory input), Facebook's algorithms may determine that men are predominately clicking on a particular advertisement and may therefore factor gender into the relevance score, producing a discriminatory outcome wherein women are completely excluded from viewing the ad (Andrew, interview, October 5, 2018). In this manner,

Facebook can override the preferences of an advertiser and cause an ad campaign to discriminate even when this was not the original intention (Ali et al. 2019).

Advertising discrimination can therefore occur in myriad ways on Facebook, several of which are covert and algorithmically based. This level of uncertainty benefits the platform by allowing it to deflect responsibility for any resulting discrimination. Troublingly, although Facebook agreed as part of a settlement in 2019 to end targeting based on gender, age, religion, race, ethnicity, or zip code for housing, credit, and employment advertisements (Jan and Dwoskin 2019), the site did not completely stop its algorithm from relying on proxy variables (for protected classes) when creating lookalike audiences (Kofman and Tobin 2019) or from utilizing problematic ad optimization and delivery processes (Ali et al. 2019). Consequently, discrimination has been allowed to continue and expand, harming users and revealing the deeply rooted problems of self-regulation.

SOCIO-ECONOMIC IMPACTS OF AD DISCRIMINATION

Although it is impossible to know (partially because of Facebook's intentional lack of transparency) how many individuals have been denied economic opportunities due to housing, credit, or employment advertising discrimination, this problematic practice has likely been impactful because of Facebook's central role in the advertising ecosystem. Facebook alone accounts for over 20 percent of the global ad market (Bernard 2018) and is used by 93 percent of social media advertisers (Donnelly 2018). Given this tremendous scope, Facebook "has an unfathomable capacity to make workers aware of economic opportunities" ("First amended class and collective action complaint" 2018, 6). However, such a capacity can also be used to shut users out of opportunities and to deny them the ability to improve their financial and social standing. In fact, online advertising discrimination maybe even more problematic than offline discrimination given Facebook's command over housing, employment, and credit advertisements. To this point, lawyer and sociologist Ifeoma Ajunwa declared: "You could even argue that [advertising discrimination] using platforms [. . .] is worse, because it's more solidified; there's no wiggle room, there's no accidental meetings" (quoted in Eidelson 2018). Thus, the very traits that make Facebook appealing to advertisers, employers, and companies (i.e., tremendous scale, large user base, detailed information about individuals, and deployment of sophisticated algorithms) make this platform a potent vehicle of discrimination and a dangerous driver of inequality.

The economic and social harms resulting from advertising discrimination compound the precarity of historically marginalized groups and communities, especially along intersectional lines ("First amended class and collective action complaint" 2018). Biased targeting practices can funnel individuals into different jobs and neighborhoods by showing them disparate messages. Algorithmic advertising discrimination may thus "artificially depress[]" ("First amended class and collective action complaint" 2018, 32) the proportion of individuals from historically marginalized groups in applicant pools for employment, housing, and credit opportunities. These online practices compound offline racial-, gender-, and age-based gaps in homeownership, employment, wages, and access to credit (e.g., Deere and Doss 2006), which are expanding due to the COVID-19 pandemic (Barone 2020; Gupta 2020). Thus, the drastic material consequences of discriminatory commercial algorithms highlight the need for robust regulatory efforts.

THE PITFALLS OF FACEBOOK'S SELF-REGULATORY SYSTEM

Given the dramatic potential consequences of discriminatory advertising, one may wonder why this practice has been tolerated on Facebook. Although many factors have led to the expansion and intensification of online advertising discrimination, the self-regulatory governance of platforms is a paramount factor. In particular, structural failings inherent to self-regulation have engendered and exacerbated advertising discrimination on Facebook through lack of transparency, lack of incentive to enact change, and lack of external oversight.

Lack of Transparency

Although billions of people throughout the world are exposed to advertisements on Facebook, only an infinitesimal number of people are privy to the inside workings of the platform's advertising system and commercial algorithms. Users are often unaware of how their data are being used, why they are receiving particular advertisements, and whether they are being discriminated against (Kim 2020). In this manner, Facebook's lack of transparency is inherently one-sided; while users are prompted and encouraged to reveal intimate details about themselves, the site offers little information about its algorithms and practices to the public (Rieke and Bogen 2018). On a societal level, the lack of transparency undermines efforts to cultivate accountability and prevents meaningful oversight.

Nearly every interviewee criticized the "information asymmetry" (Balkin 2016, 1223) between the platform and its users. The one allowance for users is the "Why am I seeing this ad" feature, which was created by Facebook to address complaints about its lack of transparency. However, the site's explanations are *"incomplete* and sometimes *misleading"* (Andreou et al. 2018, 11; italics in original). The "why am I seeing this ad" tool only reveals the most common characteristic among those who are targeted, and therefore allows advertisers to hide their discriminatory targeting practices by adding a prevalent attribute (in addition to a discriminatory one) at the time of audience selection (Andreou et al. 2018). Furthermore, the utility of this tool is questionable since algorithmic transparency does not necessarily impart an understanding of the code and the assumptions that go into constructing the algorithm in the first place (Ananny and Crawford 2018). Caleb, a privacy law professor, noted that information about targeting parameters is minimally useful to users since they do not have a way to change how the platform functions and what its business model entails (interview, November 1, 2018). Therefore, the creation of the "Why am I seeing this ad" feature and its purposeful exclusion of relevant information allow Facebook to give the appearance of being open while still withholding vital data. This opacity reveals the platform's commodification of users as it treats individuals as data reservoirs undeserving of full explanation.

Facebook's lack of transparency intentionally blocks robust research and investigatory efforts. Interviewees explained that Facebook's advertising operations exist "below the surface" (Andrew, interview, October 5, 2018) and characterized the system as a "black box" (Mason, interview, October 23, 2018), echoing how scholars conceptualize this issue (Pasquale 2015).

Consequently, researchers have struggled to devise methodologies to study Facebook's secretive advertising system. Alex, a computer science professor, explained that researchers try to parse through the "huge number of unknowns" (interview, October 23, 2018) by accessing individual accounts (through voluntarily installed browser extensions), enabling researchers to collect the advertisements users see. However, this method requires a large sample size to establish statistical confidence, meaning researchers often have to turn to platforms, such as Amazon Mechanical Turk or Prolific to recruit individuals, which may present ethical dilemmas and challenges with establishing external validity (Sheehan 2018). Andrew, a technology and social justice researcher, further explained the limitations of accessing individual profiles as researchers using this approach "can't really get the bird's-eye view on Facebook" (interview, October 5, 2018), and therefore have trouble drawing conclusions that can be used to hold the site accountable.

Balancing its corporate reputation and capitalist motives, Facebook regularly weighs its interest in avoiding public criticism against a perceived need to safeguard its technology. As explained by Chase, a marketer, since algorithmic targeting represents the company's "bread and butter" (interview,

September 26, 2018), releasing details about this system raises a host of concerns for Facebook. Mason, a privacy and security researcher, noted that Facebook worries "losing [its] trade secrets" would risk the possibility of "advertisers and others gaming [its] systems" as well as "competitors using [its] algorithms to improve their own systems" (interview, October 23, 2018). While these concerns are understandable, there are ways for Facebook to be more forthcoming about its algorithms while still ensuring the integrity of its technology. For example, Facebook could release advertising campaign metadata that would provide details about targeting options set by advertisers and demographic information about which users viewed particular ads, allowing the platform to be held accountable to a greater extent than currently possible (Rieke and Bogen 2018). In failing to release such information, Facebook places its bottom line above protection of users and compliance with federal laws.

In addition to safeguarding its revenue, Facebook's lack of transparency stonewalls legal actions. As previously mentioned, it is illegal for a site to "print or cause [a discriminatory advertisement] to be printed" (Civil Rights Act of 1964, § 2000e-3), and a platform can lose its immunity under § 230 of the CDA if it "materially contribute[s]" to a discriminatory advertisement (*Fair Housing Council of San Fernando Valley v. Roommates.com*, 521 F.3d 1157 (9th Cir. 2008)). Insights into the workings of Facebook's advertising system are necessary since there are many potential sources of advertising discrimination, including the deliberate choices of advertisers, the targeting of another advertiser who is competing in the same auction, and the decisions made by Facebook's algorithms (Lambrecht and Tucker 2019). Interviewees stressed that, by withholding details about its advertising system, Facebook disables courts from having full access to relevant information while adjudicating charges of discrimination (Gwen, interview, September 28, 2018; Orlando, interview, October 15, 2018). The nuanced aspects of Facebook's advertising system may only be fully known to a court in the event that the site is forced to disclose information, perpetuating discrimination in the interim.

The relationship between Facebook's self-regulation and its strategic opacity are mutually reinforcing to the detriment to users. The historic promotion of self-regulation in the United States (Campbell 1998) and the weak climate of external oversight enable companies to act in a nontransparent manner without fear of significant financial or political repercussions. Furthermore, Facebook's refusal to disclose information takes advantage of the immunity afforded under § 230 and strengthens the site's self-regulatory governance. As the platform insulates itself from investigatory research and legal inquiry, it leaves external regulators with no ammunition, thereby closing off the possibility of outside governance. Thus, since Facebook remains the single entity with access to its advertising algorithms, it positions itself as the only body that can regulate and revise its system, perpetuating weakly enforced self-regulation and preventing necessary accountability and transparency.

Lack of Incentive to Enact Change

Facebook's self-regulatory governance and capitalist directive powerfully contribute to the site's lack of incentive to enact change. Due to the site's economic configuration, it is constantly "straddling two markets—the social media market on the one hand and the advertising market on the other" (Orlando, interview, October 15, 2018). To deliver users to advertisers, Facebook must act as an "attention merchant" (Wu 2017, 338), keeping users on the platform long enough to be exposed to advertisements, creating a system of divided loyalties.

Caleb asserted that the company's tight-knit relationship with advertisers often comes at users' expense, whose data comprise "the life blood of [Facebook's] whole advertising system" (interview, November 1, 2018). Users are harmed by this system since Facebook's business model mandates discrimination and algorithmic targeting. To this point, Ben, a marketing professor, noted that Facebook is incentivized to enable algorithmic micro-targeting because advertisers typically pay more to reach a targeted audience (interview, October 19, 2018). Further describing the site's imperative to engage in sophisticated targeting practices, Andrew explained that payment per click is suspected to be Facebook's most common billing trigger, meaning the platform is financially motivated to put advertisements in front of the users its algorithms determine are most likely to click on them (interview, October 5, 2018). Consequently, Facebook has an "economic incentive [. . .] to do even further optimization of the delivery of ads even after the advertiser does their targeting choices" (Andrew, interview, October 5, 2018). Since hyper-segmented targeting is incredibly profitable for the site and is perceived to be effective, Facebook is motivated to ignore abuses by advertisers, and advertisers in turn are motivated to disregard dubious legal or ethical decisions on the site.

Facebook's symbiotic (and lucrative) relationship with advertisers gives the site little incentive to make sure advertisers comply with its rules. In 2017, when it was besieged by negative press, Facebook established a policy requiring advertisers to self-certify that their ads were not discriminatory (Bala 2017). While instituting this change, Facebook failed to install any measures to enforce advertisers' compliance. Not surprisingly, Facebook, a self-regulating entity, has not meaningfully monitored advertisers who are entrusted to self-regulate on the site. These multiple, nested layers of self-regulation compound problems with accountability by enabling two entities (i.e., the platform and advertisers) with mutually reinforcing capitalist aims to operate without meaningful external oversight. Since Facebook's primary source of revenue is advertisements (Fuchs 2012), the self-certification process does not encourage the platform to proactively monitor abuse due to this glaring conflict of interest. Although interviewees believed that self-certification by advertisers may be a step in the right direction, they were quick to point out that this measure is unlikely to improve the situation by itself. Orlando, a law professor with expertise in communication and information policies, critiqued the self-certification process: "Most people are not going to attest to being discriminating [. . .] and even with these protections [advertisers] are still breaking the law" (interview, October 15, 2018). Expressing similar concerns, Alex indicated the need for external oversight to provide a check on self-regulatory measures: "This is a case where you actually need enforcement rather than blindly trusting the advertisers" (interview, October 23, 2018). By refusing to actively enforce its policies, Facebook paves the way for additional harms and inequitable outcomes on the platform.

To date, Facebook's efforts to improve its advertising system have been prompted by looming external threats rather than internal directives. Citing the platform's history of promising change and failing to follow through, interviewees were skeptical of the platform's genuine commitment to reform. Orlando explained:

> While they have been improving in a lot of regards, it isn't because they're doing it necessarily out of some kind of original feeling of what is right. They're doing it because they've been pressured to do it. They're doing it because they're getting sued. (interview, October 15, 2018)

Facebook's focus on avoiding negative press and lawsuits has led it to engage in arbitrary actions that reveal the extent of its self-interest. A prime example of this tendency is Facebook's initial decision to end ethnic affinity targeting (a measure that was only somewhat successful) after *ProPublica*'s first report, but not to curb gender- or age-based targeting, which have received less press coverage (Egan 2016). According to Andrew, this selective ban does not have to do with the technical feasibility of prohibiting age- or gender-based targeting as much as the company's perception that racism would draw more public ire than sexism or ageism, which could impact the company's bottom line (interview, October 5, 2018). While Facebook immediately issued an apology after reports surfaced of discriminatory advertising based on race, the company defended discriminatory advertising based on age, justifying it as standard industry practice (Goldman 2017). Responding to the fact that Facebook prohibited race-based targeting but not gender-based targeting for employment advertisements, civil rights lawyer Gwen declared: "[It's] kind of inexplicable that they would take that step with respect to race and not address the other legal violations at play" (Gwen, interview, September 28, 2018).

A powerful example of Facebook's reactive behavior and lack of incentive to enact change can be seen in its 2019 settlement, the deficiencies of which have been mentioned in this chapter. As previously noted, the settlement failed to address Facebook's optimization procedures in which algorithms decide whom to deliver ads to. Consequently, Facebook "has not committed to making any changes to its algorithms [. . .] there is still no mechanism to ensure that Facebook isn't discriminating when it delivers the ad" (Levy 2019, para. 11). It is therefore apparent that Facebook made cosmetic changes rather than changing some of the structural problems at the heart of its advertising system. Thus, Facebook's piecemeal and reactive reforms highlight another pitfall of self-regulating platforms that lack an incentive to enact change and are empowered to apply their own set of standards and enforcement mechanisms. These elements are used selectively to maintain economic interests rather than protect users, exacerbating power differentials.

Lack of External Oversight

While Facebook may be criticized for practices that fuel discriminatory advertising, all three branches of the U.S. government perpetuate this problem through their laissez-faire approach to protecting civil liberties and monitoring self-regulation, especially in the digital realm. Although the United States is unique in the fact that it has three federal laws that prohibit discrimination in housing, credit, and employment advertising, the government has largely ignored violations of these rights when they occur on social media platforms (Harrison 2020). Such a situation has led to a state of "negotiated non-compliance" (Gunningham 1987, 91). This setup can be characterized by "a complete withdrawal from enforcement activity, a toothless, passive and acquiescent approach which has tragic consequences for those whom the legislation is ostensibly intended to protect" (Gunningham 1987, 91). In the case of advertising discrimination, both Facebook and the U.S. government have worked in concert to maintain a lax regulatory regime that upholds capitalist interests while violating the civil rights of users and compounding societal inequities.

The legislative branch has contributed to advertising discrimination through inadequate provisions and safeguards. The European Union's General Data Protection Regulation harmonizes privacy laws across member states and creates a right to explanation wherein users can request information about algorithmic decisions pertaining to them. In contrast, the United States has "piecemeal legislation" as well as "regulatory and compliance gaps"

(Caleb, interview, November 1, 2018) with respect to privacy and security online. Additionally, the United States does not have a posteriori regulation of algorithms that would mandate transparency and auditing (Kaminski 2018). Further compounding this problem, many existing legal protections for civil rights are outdated in the digital era (Allen 2019). Raymond, a law professor with expertise in housing issues, noted that both the Civil Rights Act of 1968 and the CDA (1996) are out of date and that the creators of these laws could not have possibly foreseen the development of discriminatory practices and algorithms online (interview, October 10, 2018). To this point, interviewees explained that technological developments (such as sophisticated algorithms) that facilitate discrimination evolve rapidly while the legal system is slow to make changes and fundamentally reactive (Bennett Moses 2011). Consequently, the law is perpetually playing catch-up and engaging in an endless game of "whack-a-mole" as new problems continually arise (Raymond, interview, October 10, 2018; Alex, interview, October 23, 2018).

The outdated nature of these laws is most problematic with regard to civil rights protections. Many antidiscrimination statutes lack an intersectional perspective, which limits the number of protected classes who are covered and treats discrimination against all of these groups equally (Mann and Matzner 2019). An additional problem is that many federal antidiscrimination laws were written during the Civil Rights Movement, an era when housing, employment, and credit searches occurred in-person and discrimination was overt (Allen 2019). However, more than five decades after these laws were passed, they remain almost in their original form. For instance, the Civil Rights Act of 1968 has been amended infrequently, and the advertisement provision of this statute has never been updated since its passage in 1968 except for the inclusion of additional protected classes (Spinks 2019). The massive changes in the technological landscape between the passage of the law and the present day render the statute ineffectual in preventing most forms of contemporary discrimination. Acutely aware of the lack of external oversight and the slow pace of legal reforms, Facebook has been emboldened to push the limits of legal acceptability, enabling advertising discrimination to flourish.

In addition to the legislative branch's inactivity, the judicial branch has failed to adequately curtail discriminatory advertising. Interviewees lamented the judicial branch's "ambitious and broad protection for intermediaries" (Orlando, interview, October 15, 2018), pushing § 230 beyond its logical conclusion and granting platforms outsized protections at the expense of user rights (Citron and Wittes 2017). The material contribution standard set forth in *Roommates.com* creates such a high bar to being considered a co-developer of content that companies can easily argue they have not substantively contributed to the development of an advertisement, thereby preserving their immunity (Sylvain 2018). Interviewee Orlando explained that social media companies have been protected "because courts have been taken by the romance of a free and unvarnished [space for] online discourse . . . [and] information distribution" (interview, October 15, 2018). The extent of judicial protection and lack of legal repercussions for social media companies perpetuate platforms' sense of exceptionalism and separatism, fueling their disregard for the spirit—if not the letter—of U.S. laws.

The enforcement branch, with arguably the most power to restrict discriminatory advertising, has also taken a back seat, enabling social media companies to continue harmful practices without effective oversight. Although one of HUD's primary objectives is to prevent housing discrimination (HUD, n.d.), Raymond insisted: "We currently have a federal government that in almost every way has no interest in the Fair Housing Act" (interview, October 10, 2018). HUD's efforts to ensure compliance with the Civil Rights Act of 1968 have declined in the current administration due in part to a reduced focus on racial justice (Thrush 2018), a trend that has benefited Facebook. Shortly after HUD opened an investigation of the company in 2016, Facebook's half-hearted promise to initiate reforms convinced many government officials to "back off," conveniently believing the problem would soon be solved without their having to directly intervene or sanction the company (Raymond, interview, October 10, 2018). In 2017, HUD closed its investigation without an explanation. It should be noted that HUD resumed its investigation of Facebook's discriminatory advertising practices after *The New York Times* published an exposé in 2018 about HUD's lack of fair housing enforcement (Thrush 2018). Furthermore, HUD filed charges against the platform several months before the 2019 U.S. elections, "a time when conservatives [were] ramping up pressure on big tech platforms" as part of their political agenda (Newton 2019, para. 14). This timing suggests that HUD was not motivated to examine Facebook's abuses until the agency came under press scrutiny and saw a political opportunity, self-interested actions that perpetuated discriminatory advertising for years.

CONCLUSION: ALGORITHMS, SELF-REGULATION, AND ADVERTISING DISCRIMINATION

In many ways, algorithmic advertising discrimination is a microcosm of larger intersecting debates about the role of platforms in democratic society, the capacity of algorithms and machine learning to either mitigate or deepen social inequities, and the pitfalls of self-regulatory governance. The emergence of social media sites and the increased reliance on algorithms were initially greeted with optimism, utopianism, and almost religious fervor, heralded as the great deliverers of equality, democracy, and liberation (Shapiro 1999). As companies spoke of a teleological thrust of progress, many individuals, businesses, and governments became swept up in this grand vision, eagerly investing in platforms and algorithms and refusing to meaningfully regulate them. While sites such as Facebook and the sophisticated commercial algorithms they employ have brought society many benefits, we have also paid a high price as discrimination, inequity, and exclusion are the sine qua non of these technologies. For too long, we have allowed social media sites to self-regulate, empowering platforms to prioritize corporate profitability over user equality.

While installing self-regulation may have made sense in the 1990s when the commercial Internet was still in its infancy, it is hard to justify this regulatory approach decades later when the misdeeds of social media sites and the tangible harms their algorithms have triggered have become all too apparent (Citron and Wittes 2017). In fact, structural elements indicate why self-regulation is an inappropriate method for curbing the reckless behavior of platforms, advertisers, and even government agencies who have all functioned with limited oversight. The social media market is immature, there is no established tradition of cooperation with state authorities, and most social media sites have not previously exhibited sensitivity to public interests (Saurwein 2011). Additionally, the business model of platforms does not incentivize meaningful self-regulation as there is a fundamental conflict between public and private interests since the very features (e.g., harvesting of user data to refine algorithmic predictions) that make these sites profitable also facilitate discrimination. Most significantly, self-regulation is an unacceptable form of governance when applied to media or technology (including algorithms) that have the capacity to discriminate against individuals (especially those from historically marginalized groups) and trigger material offline consequences, such as the abridgment of civil rights and the curtailment of economic opportunities.

It is critically important to reevaluate our regulatory paradigm for platforms and commercial algorithms, drawing upon both structural analyses and qualitative research with the groups most impacted by self-regulatory practices. Scholars need to meaningfully engage with the question of regulation, moving beyond arguments that algorithms should not be regulated (Reed 2018) or that value-laden inputs, which are unlikely to ever be exposed, should form the basis of regulation (Noble 2018). Furthermore, scholars and policymakers should focus their attention on both a priori measures (i.e., before algorithms are deployed) such as ethical design solutions and a posteriori regulations of algorithms (i.e., regulation after algorithms are deployed) such as external transparency, reporting, and auditing requirements. When combined, these approaches may lead to lasting changes, tackling the structural roots of self-regulatory governance and opening systems up to outside review.

Concomitantly, regulators should follow the examples set by countries and blocs around the world who are pursuing more co-regulatory or command-and-control approaches to constrain social media sites while the United States stubbornly clings to a self-regulatory model (Nurik 2019). In doing so, the government grants platforms a sense of exceptionalism that has enabled them to wall themselves off from U.S. laws and to engage in unethical (and in some cases, even illegal) practices, hiding behind the opacity of their algorithms. Specifically, Facebook's elaborate advertising system built around complex algorithms and persistent user segmentation, in combination with its lack of transparency, lack of incentive to enact change, and lack of external oversight, have abetted and enabled potent advertising discrimination. These elements have combined to create the perfect storm: a platform that answers to shareholders instead of the public, keeps its algorithms and even its transgressions secret, and functions with almost complete immunity away from the watchful eye of the government. It is only when structural changes are implemented, when self-regulation is curbed, and when sites and the algorithms they profit from are regarded with skepticism instead of blind trust that we can begin addressing and reforming the discriminatory patterns endemic to social media giants.

NOTES

1. For one attempt to get around this secrecy and reconstruct Facebook advertising and weblining practices through experiment, see Sawchuk, DeJong, and Gauthier, chapter 2, this book. — Ed.

2. On the instantaneity and temporality of algorithms, and how encounters with machine learning or this nature might be allegorized in fiction, see Cox, chapter 11, this book. — Ed.

REFERENCES

Age Discrimination in Employment Act of 1967, 29 U.S.C. § 623.

Ali, Muhammad, Piotr Sapiezynski, Miranda Bogen, Aleksandra Korolova, Alan Mislove, and Aaron Rieke. 2019. "Discrimination through Optimization: How Facebook's Ad Delivery Can Lead to Biased Outcomes." *Proceedings of the ACM on Human-Computer Interaction*, 3 (CSCW): 1–30.

Allen, James A. 2019. "The Color of Algorithms: An Analysis and Proposed Research Agenda for Deterring Algorithmic Redlining." *Fordham Urban Law Journal*, 46: 219–270.

Ananny, Mike, and Kate Crawford. 2018. "Seeing Without Knowing: Limitations of the Transparency Ideal and Its Application to Algorithmic Accountability." *New Media & Society*, 20 (3): 973–989.

Andreou, Athanasios, Giridhari Venkatadri, Oana Goga, Krishna P. Gummadi, Patrick Loiseau, and Alan Mislove. 2018. "Investigating Ad Transparency Mechanisms in Social Media: A Case Study of Facebook's Explanations." *The Network and Distributed System Security Symposium (NDSS)*. https://doi.org/10.14722/ndss.2018.23204

Angwin, Julia, and Terry Parris. 2016. "Facebook Lets Advertisers Exclude Users by Race." *ProPublica*, October 28, 2016. https://www.propublica.org/article/facebook-lets-advertisers-exclude-users-by-race

Angwin, Julia, Noam Scheiber, and Ariana Tobin. 2017. "Dozens of Companies are Using Facebook to Exclude Older Workers from Job Ads." *ProPublica*, December 20, 2017. https://www.propublica.org/article/facebook-ads-age-discrimination-targeting

Baker, C. Edwin. 1994. *Advertising and a Democratic Press*. Princeton: Princeton University Press.

Bala, Divakar. 2017. "Targeting Exclusions-Update." *Facebook*, December 19, 2017. https://developers.facebook.com/ads/blog/post/2017/12/19/targeting-exclusions-update-blog-post/

Balkin, Jack M. 2016. "Information Fiduciaries and the First Amendment." *UCDL Review*, 49 (4): 1183–1234.

Barone, Emily. 2020. "The Housing Market is Booming, but Millions Face Eviction—And the Gap is Getting Worse." *Time*, December 4, 2020. https://time.com/5917894/evictions-housing-market-covid/

Bennett Moses, Lyria. 2011. "Agents of Change: How the Law 'Copes' with Technological Change." *Griffith Law Review*, 20 (4): 763–794.

Bernard, Zoë. 2018. "Facebook Accounts for 20% of the Global Online Ad Market." *Business Insider*, March 28, 2018. https://www.businessinsider.com/facebook-advertising-market-share-chart-2018-3

Beyers, Jan, Caelesta Braun, David Marshall, and Iskander De Bruycker. 2014. "Let's Talk! On the Practice and Method of Interviewing Policy Experts." *Interest Groups & Advocacy*, 3 (2): 174–187.

Bozdag, Engin, and Jeroen van den Hoven. 2015. "Breaking the Filter Bubble: Democracy and Design." *Ethics and Information Technology*, 17 (4): 249–265.

Campbell, Angela J. 1998. "Self-Regulation and the Media." *Federal Communications Law Journal*, 51 (3): 711–772.

Charge of Discrimination. *ACLU, Outten & Golden LLP, and the Communications Workers of America v. Facebook*, 2018. Retrieved from https://www.aclu.org/legal-document/facebook-eeoc-complaint-charge-discrimination

Chen, Shu-Heng. 2012. *Genetic Algorithms and Genetic Programming in Computational Finance*. New York: Springer Science & Business Media.

Citron, Danielle Keats, and Benjamin Wittes. 2017. "The Internet Will Not Break: Denying Bad Samaritans Sec. 230 Immunity." *Fordham Law Review*, 86: 401–423.

Civil Rights Act of 1964, 42 U.S.C. § 2000e-3.

Civil Rights Act of 1968, 7 CFR § 1901.203(b)(3)(ii).

Collins, Stephen. 2008. "Saving Fair Housing on the Internet: The Case for Amending the Communications Decency Act." *Northwestern University Law Review*, 102 (3): 1471–1500.

Communications Decency Act of 1996, 47 U.S.C. § 230.

Danna, Anthony, and Oscar H. Gandy. 2002. "All that Glitters Is Not Gold: Digging Beneath the Surface of Data Mining." *Journal of Business Ethics*, 40 (4): 373–386.

Datta, Amit, Arunpam Datta, Jael Makagon, Deirde K. Mulligan, and Michael Tschantz. 2018. "Discrimination in Online Advertising: A Multidisciplinary Inquiry." *Conference on Fairness, Accountability and Transparency*, 20–34.

Deere, Carmen Diana, and Cheryl R. Doss. 2006. "The Gender Asset Gap: What Do We Know and Why Does It Matter?" *Feminist Economics*, 12 (1–2): 1–50.

DeVito, Michael A. 2017. "From Editors to Algorithms: A Values-Based Approach to Understanding Story Selection in the Facebook News Feed." *Digital Journalism*, 5 (6): 753–773.

Donnelly, Gordon. 2018. "75 Super-Useful Facebook Statistics for 2018." *WordStream*, September 7, 2018. https://www.wordstream.com/blog/ws/2017/11/07/facebook-statistics

Eidelson, Josh. 2018. "Facebook Tools Are Used to Screen out Older Job Seekers, Lawsuit Claims." *Bloomberg*, May 28, 2018. https://www.bloomberg.com/news/articles/2018-05-29/facebook-tools-are-used-to-screen-out-older-job-seekers-lawsuit-claims.

Egan, Erin. 2016. "Improving Enforcement and Promoting Diversity: Updates to Ethnic Affinity Marketing." *Facebook*, November 11, 2016. https://newsroom.fb.com/news/2016/11/updates-to-ethnic-affinity-marketing/

Equal Credit Opportunity Act of 1974, 15 U.S.C. § 1691.

Eubanks, Virginia. 2017. *Automating Inequality: How High-Tech Tools Profile, Police, and Punish the Poor.* New York: St. Martin's Press.

Facebook. n.d. (a). "About the Delivery System: Ad Auctions." Accessed November 1, 2018. https://www.facebook.com/business/help/430291176997542

———. n.d. (b). "Choose your Audience." Accessed November 1, 2018. https://www.facebook.com/business/products/ads/ad-targeting

———. n.d. (c). "Core Audiences." Accessed November 1, 2018. https://www.facebook.com/business/products/ads/ad-targeting#core_audiences

———. n.d. (d). "Reach the Right People." Accessed November 1, 2018. https://www.facebook.com/business/m/interest-targeting

Fair Housing Council of San Fernando Valley v. Roommates.com, 521 F.3d 1157 (9th Cir. 2008).

Ferrer-Conill, Raul, Erik Knudsen, Corinna Lauerer, and Aviv Barnoy. 2020. "The Visual Boundaries of Journalism: Native Advertising and the Convergence of Editorial and Commercial Content." *Digital Journalism*. https://doi.org/10.1080/21670811.2020.1836980

First Amended Class and Collective Action Complaint. *Bradley v. T-Mobile US, Inc.*, 2018. Retrieved from https://www.courtlistener.com/docket/6247542/56/bradley-v-t-mobile-us-inc/

First Amended Complaint. *Onuoha v. Facebook*, 2017. Retrieved from https://www.courtlistener.com/docket/4496060/28/onuoha-v-facebook-inc/

Fuchs, Christian. 2012. "The Political Economy of Privacy on Facebook." *Television & New Media*, 13 (2): 139–159.

Gunningham, Neil. 1987. "Negotiated Non-Compliance: A Case Study of Regulatory Failure." *Law & Policy*, 9 (1): 69–95.

Gupta, Alisha Haridasani. 2020. "Why Did Hundreds of Thousands of Women Drop out of the Work Force?" *The New York Times*, October 3, 2020. https://www.ny times.com/2020/10/03/us/jobs-women-dropping-out-workforce-wage-gap-gender .html

Harrison, Dominique. 2020. "Civil Rights Violations in the Face of Technological Change." *Aspen Institute*, October 22, 2020. https://www.aspeninstitute.org/blog -posts/civil-rights-violations-in-the-face-of-technological-change/

HUD. n.d. "General Information about HUD Contracting." Accessed November 1, 2018. https://www.hud.gov/program_offices/sdb/guide/general#HUDdo

Jan, Tracy, and Elizabeth Dwoskin. 2019. "Facebook Agrees to Overhaul Targeted Advertising System for Job, Housing and Loan Ads after Discrimination Complaints." *The Washington Post*, March 19, 2019. https://www.washingt onpost.com/business/economy/facebook-agrees-to-dismantle-targeted-advertising -system-for-job-housing-and-loan-ads-after-discrimination-complaints/2019/03/19 /7dc9b5fa-4983-11e9-b79a-961983b7e0cd_story.html?noredirect=on&utm_term= .469b532bf069

Lambrecht, Anja, and Catherine Tucker. 2019. "Algorithmic Bias? An Empirical Study of Apparent Gender-Based Discrimination in the Display of STEM Career Ads." *Management Science*, 65 (7): 2966–2981.

Levy, Pema. 2019. "Facebook is Cracking Down on Ad Discrimination. But the Bias May Be Embedded in Its Own Algorithms." *Mother Jones*, July 12, 2019. https ://www.motherjones.com/politics/2019/07/facebook-is-cracking-down-on-ad-dis crimination-but-the-bias-may-be-embedded-in-its-own-algorithms/

Kaminski, Margot E. 2018. "Binary Governance: Lessons from the GDPR's Approach to Algorithmic Accountability." *Southern California Law Review*, 92: 1529–1616.

Kim, Pauline. 2020. "Manipulating Opportunity." *Virginia Law Review*, 106: 867–935.

Kofman, Ava, and Ariana Tobin. 2019. "Facebook Ads Can Still Discriminate Against Women and Older Workers, Despite a Civil Rights Settlement." *ProPublica*, December 13, 2019. https://www.propublica.org/article/facebook -ads-can-still-discriminate-against-women-and-older-workers-despite-a-civil -rights-settlement

Mann, Monique, and Tobias Matzner. 2019. "Challenging Algorithmic Profiling: The Limits of Data Protection and Anti-Discrimination in Responding to Emergent Discrimination." *Big Data & Society*, 6 (2). https://doi.org/10.1177 /2053951719895805

Martin, Kirsten. 2018. "Ethical Implications and Accountability of Algorithms." *Journal of Business Ethics*. https://doi.org/10.1007/s10551-018-3921-3.

Murphy, Erin. 2007. "The New Forensics: Criminal Justice, False Certainty, and the Second Generation of Scientific Evidence." *California Law Review*, 95 (4): 721–798.

Newton, Casey. 2019. "Why HUD's Lawsuit Against Facebook Came as a Surprise." *The Verge*, March 29, 2019. https://www.theverge.com/interface/2019/3/29/18286 088/hud-facebook-lawsuit-ben-carson-fair-housing

Noble, Safiya. 2018. *Algorithms of Oppression: How Search Engines Reinforce Racism*. New York: New York University Press.

Nurik, Chloé. 2018. "50 Shades of Film Censorship: Gender Bias from the Hays Code to MPAA Ratings." *Communication, Culture and Critique*, 11 (4): 530–547.

———. 2019. "'Men are Scum': Self-Regulation, Hate Speech, and Gender-Based Censorship on Facebook." *International Journal of Communication*, 13: 2878–2898.

Pasquale, Frank. 2015. *The Black Box Society: The Secret Algorithms that Control Money and Information*. Cambridge: Harvard University Press.

Reed, C. 2018. How Should We Regulate Artificial Intelligence? *Philosophical Transactions of the Royal Society A: Mathematical, Physical and Engineering Sciences*, 376 (2128): 20170360.

Rieke, Aaron, and Miranda Bogen. 2018. "Leveling the Platform: Real Transparency for Paid Messages on Facebook." *Upturn*. https://www.upturn.org/static/reports/20 18/facebook-ads/files/Upturn-Facebook-Ads-2018-05-08.pdf

Saurwein, Florian. 2011. "Regulatory Choice for Alternative Modes of Regulation: How Context Matters." *Law & Policy*, 33 (3): 334–366.

Scheiber, Noam. 2018. "Facebook Accused of Allowing Bias Against Women in Job Ads." *The New York Times*, September 18, 2018. https://www.nytimes.com/2018 /09/18/business/economy/facebook-job-ads.html

Shapiro, Andrew L. 1999. *The Control Revolution: How the Internet is Putting Individuals in Charge and Changing the World We Know*. New York: Public Affairs.

Sheehan, Kim Bartel. 2018. "Crowdsourcing Research: Data Collection with Amazon's Mechanical Turk." *Communication Monographs*, 85 (1): 140–156.

Smythe, Dallas. 1977. "Communications: Blindspot of Western Marxism." *Canadian Journal of Political and Social Theory*, 1 (3): 1–27.

Spinks, Chandler Nicholle. 2019. "Contemporary Housing Discrimination: Facebook, Targeted Advertising, and the Fair Housing Act." *Houston Law Review*, 57 (4). https://houstonlawreview.org/article/12762-contemporary-housing-discrimination -facebook-targeted-advertising-and-the-fair-housing-act

Spurgeon, Christina. 2007. *Advertising and New Media*. London: Routledge.

Speicher, Till, Muhammad Ali, Giridhari Venkatadri, Felipe Nunes Ribeiro, George Arvanitakis, Fabrício Benevenuto, Krishna P. Gummadi, Patrick Loiseau, and Alan Mislove. 2018. "Potential for Discrimination in Online Targeted Advertising." *Conference on Fairness, Accountability and Transparency*, 5–19.

Sweeney, Latanya. 2013. "Discrimination in Online Ad Delivery." *Queue*, 11, no. 3: 10–29.

Sylvain, Olivier. 2018. "Intermediary Design Duties." *Connecticut Law Review*, 50 (1): 203–277.

Thrush, G. 2018. "Under Ben Carson, HUD Scales Back Fair Housing Enforcement." *New York Times*, March 28, 2018. https://www.nytimes.com/2018/03/28/us/ben -carson-hud-fair-housing-discrimination.html

Tobin, Ariana. 2018. "Facebook Promises to Bar Advertisers from Targeting Ads by Race or Ethnicity. Again." *ProPublica*, July 25, 2018. https://www.propublica.org /article/facebook-promises-to-bar-advertisers-from-targeting-ads-by-race-or -ethnicity-again

Tufekci, Zeynep. 2015. "Facebook Said Its Algorithms Do Help Form Echo Chambers, and the Tech Press Missed It." *New Perspectives Quarterly*, 32 (3): 9–12.

Turow, Joseph. 2011. *The Daily You: How the New Advertising Industry is Defining Your Identity and Your Worth*. New Haven: Yale University Press.

Wu, Tim. 2017. *The Attention Merchants: The Epic Scramble to Get Inside Our Heads*. New York: Vintage.

Chapter 2

Faking Age?

Ageing and the Algorithmic Assemblage

Kim Sawchuk, Scott DeJong, and Maude Gautier

There is a burgeoning literature in media studies on the relationship between algorithms and identity. This includes robust discussions on how algorithms re-produce sexist and racist profiles (Noble 2016; Chun 2019; Cheney-Lippold 2011). Yet one aspect of identity that has been largely overlooked is the intersection between algorithms and age. This is ironic, for one of the most ubiquitous demographic variables is date of birth and a common question that is asked in any data collection exercise is "How old are you?"

In this chapter, we examine the nexus between algorithmic media and this thing called "age" within "algorithmic culture" (Striphas 2015). While algorithms increasingly assert an influence on us, offering continual recommendations based on calculations of past behaviors (Tondello, Orji, and Nacke 2017), they are notoriously difficult to discern as they are "blackboxed" and proprietary (Pasquale 2016; Gillespie 2016; Sandvig et al. 2016). How can researchers study algorithms that are closely guarded, IP protected, blackboxed, and difficult to reverse engineer? Within this environment, how does date of birth, or "real age," factor within the algorithmic calculus?

Using age as our research lens, we discuss a reverse-engineering experiment conducted over a six-month period in 2019, to better understand how, algorithmically speaking, age and ageing matter. We created two fictional Facebook avatars named Ur, a 72-year-old individual whose gender is unknown, and Bernadette, a 52-year-old woman. The stories of Ur and Bernadette's cyber-fate are contrasted with observations on how we, as researchers, have been targeted with messages and advertisements over this same period. By "faking age" we reveal how age-related or, more accurately, ageist images are generated within the current algorithmic media assemblage. At approximately the same time as we created these fictional identities, we conducted an inter-generational "auto-ethnography" keeping a watchful

eye on the age-related pop-ups that appeared on our social media platforms
(YouTube; Instagram; Facebook). The fictional identities were relatively
passive and did not engage in a full range of activities: they were primarily
identities with a date of birth and an ascribed gender, with no robust network of contacts or friends—and perhaps most importantly, no purchasing
behaviors. Our actual identities generated, of course, a far wider range of
age-related activities. However, in both cases, what was reproduced were
rather predictable depictions of appropriate behaviors and interests related
not only to age but to our placement within different phases—as a student,
mother, and grandmother—of the life course. As we argue, while algorithmic
media tie age to other "interests" and identity categories, including gender,
socio-economics, race, location, and sexuality, these media feed off existing
cultural representations of age and ageing. Accordingly, they discursively
re-produce a limited understanding of ageing within this new interconnected
territory of devices, social media platforms, and search engines.

WHAT IS AN ALGORITHM?

An algorithm is a set of prescriptive steps, a calculation, to be followed to
achieve a desired output based on the continual collection of inputs or data
(Kitchin 2017). While algorithms have a long history (Striphas 2015) predat-
ing the advent of computer-mediated communications and online activities,
more recently algorithms have attracted much discussion as researchers
document the impact of machine-based learning on Internet behaviors.
Commercially driven algorithmic media engage in the ongoing gathering,
sorting, and analysis of user data from a multiplicity of sources to produce
user preferences which can be used to suggest particular purchasing behaviors, influence voting patterns, streamline information to the right audience,
or profile potential terrorist activities because of one's associations (Kitchin
2017). In terms of consumerism, this way of profiling individuals to reach
potential consumers, or target audiences, is a marked shift in research derived
from approaches based on mass-marketing in the 1950s, to the development
of target marketing in the 1980s, to what the industry now describes as
interest-based advertising (Sawchuk 1996).

Algorithmic media is characterized as a computationally driven information system that engages in data gathering for predictive purposes that
can be implemented in real-time. Algorithmic media often is described as
interactional, as it is predicated on our actions and behaviors (McKelvey
2014; Cheney-Lippold 2017). As McKelvey specifies, no one algorithm is
constitutive of algorithmic media. They only have an influence when they are
running: algorithms are characterized by their responsiveness to a continual

flow of new inputs and information from not just one, but multiple users. In fact, specific ads target us based upon our similarities with other users. Algorithmic media calculations are an assemblage of interconnected varied algorithms based on both individual behaviors and those of others with whom that individual has become algorithmically associated. This impersonal yet highly personal system is what connects algorithmic media to AI, machine learning, and big data: the information gathered is not based on your individual profile alone. Age is one quick form of association.

Theoretically and politically, algorithmic media are, in Latourian terms, socio-technical assemblages which play a significant and influential role in shaping the current digital media landscape and our current image culture. Kitchin helpfully explains that socio-technical assemblages are made of "a heterogeneous set of relations including potentially thousands of individuals, data sets, objects, apparatus, elements, protocols, standards, laws, etc" (2017, 20). What is important to note is that algorithmic media as socio-technical assemblages exercise what theorists term "'soft' power" (Foucault 1971, 1976; Nye 1990). For Michel Foucault, power is subtle and contingent, intrinsically related to knowledge that constitutes individuals as objects, who come to think of themselves as subjects related to the assemblage's categories. Instead of repression, it works through normalization (Foucault 1971, 1976). Systems of soft power are not overt, and meld with digital culture through subtly pushing people in a particular direction by making assumptions of our future desires based on the continual gathering and reformulation of user data while targeting individuals as they engage across different devices and platforms. So why should we care? And how does this relate to age and ageing?

When media become algorithmic, they are comprised of a dynamic mixture of code, devices, content, significations, subjectivities, and ways of doing that can be fed back into the system immediately, in real-time. Such algorithms, as Beer notes, can produce truths, or rather "outcomes that become or reflect wider notions of truth" (2017, 8), which are operationalized through specific algorithmic deployment. As Toronto clinicians have noted in an analysis of the use of algorithms for diagnosing older patients who have lost the capacity for speech and have impaired communication, the data sets used to create the diagnostic algorithm used to assess pain thresholds are based on the data of "younger users" and facial expressions culled from a very limited dataset of mostly light-skinned participants, producing biases that can lead to treatment error (Taati et al. 2019).

Power is operationalized through data collection which uses algorithms to sort information and thus cement, maintain, or produce certain truths about lived reality. One of these realities is what it means to age in this culture. As we researched this chapter, we noted how we were being targeted as related to the differences in our ages: Kim consistently receives ads that ostensibly target her health but are predicated on body shaming: belly fat, sagging chins, and grey hair are all positioned as features to hide. Scott, as a young male in his twenties, is targeted with ads for partying and drinking. Maude, in her thirties and a mother, receives a plethora of ads about furniture for toddlers, food, ads linked to specific purchases, and professional activities. Here we see how the representations that populate our own social media are based on assumptions related to behaviors associated with what social gerontology terms the "life course."

While age is one of the key data inputs required for getting a credit card, joining a social media platform, or accessing some sites where there are legal age restrictions (for purchasing cannabis, for example), there is relatively little discussion on the specific intersections between ageing and algorithms.[1] While not explicitly about algorithms, research on quantified age and ageing explores how "big data" acts in a quasi-performative manner to ensure that ageing bodies are measurable and manageable (Katz and Marshall 2018). Katz and Marshall's research on FitBits exposes how older adults are invited to measure and prevent their bodies' decline with the help of devices such as FitBits, applications such as Brain Games, or through the aid of technologies, such as fall detectors, to stay in their homes for a longer period. This renders bodies as data, entangled in calculations of future health risk, often conflated with age-related decline.[2]

Algorithms, in other words, are part of the turn to responsive machine learning that is increasingly a part of the datafication of our inter-subjective, relational, and intersectional identities. One key performative aspect of this datafied identity is age. Not only is age-related information being used for diagnostic purposes, as Taati et al. (2019) and others have identified, but they are also being used to determine insurance rates, monitor behaviors in public, and calculate risks, such as propensity of heart disease, that could have an impact on the cost of medical insurance if one lives in a system of privatized insurance rates. As the interplay between algorithms and our lives extends beyond not just our associations but also those of others, algorithmic representations have the potential to systematically nudge personal behaviors, warranting a more careful investigation of this topic.

APPROACHES TO THE STUDY OF ALGORITHMS

Researchers have attempted a range of methodologies to explore algorithms, each providing specific insights into their processes. Striphas (2015) examines algorithmic culture by providing a historical context to understand the origins of algorithmic practices. In a series of blog posts, Sandvig et al. (2016) offers exercises to foster awareness among his students on how Facebook and

Google gather information. Noble (2016) uses keywords, supplemented by interviews with industry workers, to deconstruct how search engines racially profile young Black women through algorithms that purport to be "neutral."

We are influenced by the work of geographer Rob Kitchin as well as media theorist Fenwick McKelvey. Writing in 2014, McKelvey advocates for "democratic methods" to translate algorithms' operations into something publicly tangible, such as controversies regarding traffic shaping. McKelvey suggests that "publics offer a valuable means of generating knowledge about algorithmic media" (204, 607), and so it is that we turn to those affected by these issues, to those who have a direct stake in contributing to resolving them. It is the affected public that can best explain the consequences of algorithms and why they matter: "The investigation of algorithmic media therefore requires plural, partial, and experimental methods—democratic methods—to assist the public in their self-reflection and learning" (McKelvey 2014, 607).

According to Kitchin (2017), the main challenges of studying algorithms are three-fold. Like McKelvey, Kitchin suggests that researchers study algorithms while they are working, in interaction. Kitchin offers six approaches for researching algorithms: (1) examining pseudo-code/source code; (2) reflexively producing code; (3) reverse engineering; (4) interviewing designers or conducting an ethnography of a coding team; (5) unpacking the full socio-technical assemblage of algorithms; (6) examining how algorithms do work in the world. As Kitchin points out, some of these approaches decontextualize algorithms from the wider socio-technical assemblage in which they operate. Others require immense resources to conduct extensive fieldwork and contact coding teams. We follow both McKelvey and Kitchin's suggestions that researchers experiment on a methodological level, and we take to heart McKelvey's message. Those who are ageing are a part of "the affected public."

We are all ageing of course, but as three researchers located at different moments in what can be understood as very different life-course trajectories, we decided to use both our own experiences and to see if and how one might experiment with and disentangle how age is shaped within algorithmic media systems. We cannot turn back the clock, yet we can reverse engineer, as Kitchen suggests—but what exactly could this mean? Basically, when reverse engineering, researchers look at inputs and outputs, in our case the inputs of behavior, location, relations (e.g., clicks), and outputs of ads, sponsored, or recommended content to understand how algorithms weight and preferences some criteria and what they do. Kitchin gives the example of searching Google using the same terms on multiple computers to check if its algorithms produce different results. We first tested that: the three authors entered the same keywords (Old lady, mature women, mature men, etc.). The results were often similar but ranked in a different order: indeed, for the youngest male in our group, the suggestions were often related to his presumed sexual interest. Furthermore, the terms "mature" in front of man and woman produced a very different set of images: the men are dashing, the women often highly sexualized. What was noticeable in all cases was the predominance of whiteness when it comes to the circulation of images of "mature ageing" within social media, images that also reproduce and assume heteronormativity as a sign of successful ageing (Marshall 2017).

Bucher (2012) suggests testing Facebook's algorithm that prioritizes users' news feed by posting and interacting and examining the differences. In line with this, we created two *fictional* profiles to understand the intricacies of the intersection of age and algorithms in social media territories: mainly Facebook, but also Twitter, LinkedIn, and YouTube, and various related sites and apps. Like netnography (Kozinets 2015) and digital ethnography, part of our methodology is rooted in online participant observation. However, these are not naturally occurring in the case of the two fictional profiles that we created to interact, in premeditated ways, with the algorithms that permeate these different social media sites. Would this experiment allow us to understand, with more precision, how algorithms "produce age?" Alongside these fictional profiles, as mentioned, we decided to pay attention to how our different generational, gendered, and socio-economic positions would compare to these algorithmic profiles.

UR AGE AND BERNADETTE MCDONALD: FACEBOOK CHARACTERS

Reverse engineering is a recent innovative method that was devised specifically to address the rise of algorithms. To begin the project, we pondered the potential of how the reverse engineering method explained by Kitchin could be deployed to understand how age matters within algorithmic media systems: "By examining what data are fed into an algorithm and what output is produced it is possible to start to reverse engineer how the recipe of the algorithm is composed (how it weights and preferences some criteria) and what it does" (Kitchin 2017, 24). We decided to create these profiles not to deceive but rather because our existing profiles are marked by our years of online behaviors.

We created a profile for a fictional non-gendered character named Ur Age (Your Age) born in 1947 (age seventy-two). Ur's interests focus on topics stereotypically associated with age and ageing (such as grandparenting, retirement, a newspaper for seniors), as well as a few other groups (classical music, for example, and cycling). We first set up a gmail account, using the same name, which then allowed us to set up the Facebook profile. We used one of the authors' laptops and used different browsers. Ur's settings (profile and browser) were chosen in order to leave as many footprints as possible, because we ultimately needed Ur to get targeted by advertisers. Our experiment started on July 3, 2019, and we monitored the profile for approximately a year.

During this time, we interacted with this profile in various ways (status updates, likes, shares, comments) at least three times a week. We shared articles found on the web related to ageing (hiking for seniors, for example), and liked posts from various groups, such as Grandparents and Grandkids. We ran Google searches on topics such as knee pain, a common ailment in Kim's family. We clicked on ads and feigned interest in buying products, putting them in our basket for later purchase. We determined what to do based on the experience and research of our oldest group member (in her sixties) of activities for seniors (family, health problems, retirement, hiking for seniors, etc.) that were commonly found on her own feeds, as well as those in their seventies and eighties that were interviewed for a prior project, activities often presumed to be of interest to older adults.

We are well aware that *all* older people are not solely interested in these topics and that real people's behaviors online might lead machines to infer an age that is different than their chronological age; for instance, if one's interests focus on topics normally associated with youth.[3] By visiting sites and groups already associated, culturally, with an older age, we wanted to see if and how these profiles would be targeted based on age. We recorded Ur's online behavior (posts liked, pages visited, etc.) as well as recommended and sponsored content on Facebook, trying to see if there were connections between the two.

After creating Ur's profile, we decided to create a second profile, for a younger older adult, a woman named Bernadette McDonald, born in 1957 (age fifty-two). We used one author's desktop university PC. To complement the experiment, we decided to systematically document the ads that our personal profiles received, taking screenshots from both our computers and phones.

The content that began to appear on Ur Age's Facebook page primarily came from the groups Ur liked. What we noticed were suggested groups, as well as a list of "people you may know," most likely based on our location and the people who happened to like the same pages as Ur—people of course that Ur could never have met, as Ur does not exist except as an online entity. Suggested groups related to ageing included those trading jokes, interested in religion, and grandparenting groups. Later in the experiment, we noted that a new suggested group about grief support and another one for workshops on "life lessons" began to appear. This content associated Ur's interests with god, grief, and grandparenting. It also generated numerous offensive images

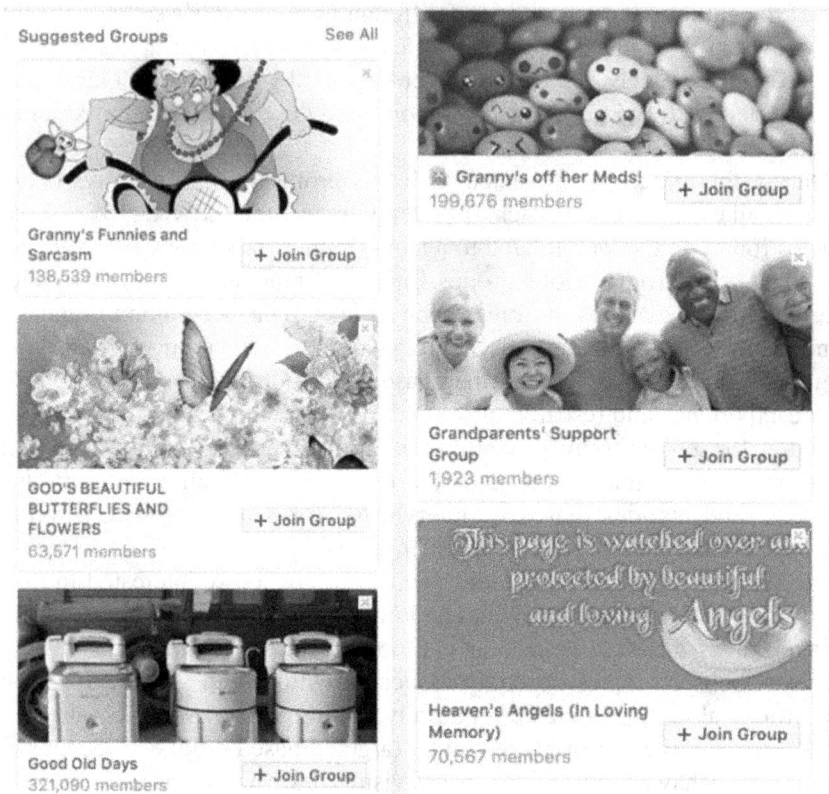

Figure 2.1   Suggested Groups.*Source*: Screen captures from Facebook by the researchers.

that associated grannies with silliness or crankiness, representing a stereo-typical, ageist figure of "the old lady" (see figure 2.1). Ur and Bernadette received numerous recommended posts (that appeared in their newsfeed), which seemed related to groups and pages they had liked, although others seem unexplainable. However, as noted, what is represented back to users is not only based on what they did, even as fictional individuals, but also what other site members do or what those whom the algorithm calculates as having similar profiles are projected as potentially doing or liking.

The second fictional profile, Bernadette McDonald, is revealing. We cre-ated Bernadette's email account using a provider (Outlook/Hotmail) that has advertisements (on the right side) and immediately received ads about life and health insurance. In the following days, however, the same ad for Quickbooks kept showing up, which does not seem related to her birth date (but might be related to Maude's digital traces). We decided to see what would happen if we searched Google for anti-ageing cream and so we clicked on an Aveeno

Recommended Post   ...

CBC Radio
July 4 at 6:00 PM · 🌐

Remember VHS tapes? A Vancouver artist is finding ways to recycle the
black spools of tape — like crocheting them into a hat.
http://www.cbc.ca/1.5194999

From the early 1970s when I discovered that the
two-inch videotape that your television programs

👍😮❤ 23   10 Comments 8 Shares

👍 Like   💬 Comment   ↪ Share

Figure 2.2   CBC Radio, Recommended Post, from Ur's Profile.*Source:* Screen capture
from Facebook by the researchers.

link. The next day, we looked up another anti-ageing cream site (Clarins)
from Google ads. Then, we went on Tena and Aveeno's Facebook pages and
clicked on links (one for their website, one on a post with the button "shop
now"). Within a matter of days, Bernadette began receiving recommended
material at a faster rate than poor Ur (see figure 2.2).

It took three weeks before we saw an ad in the sidebar—top right corner
for Ur—a Capital One Mastercard to improve Ur's credit and another from
Scotiabank, which appeared right under it. We clicked and surfed the site for
a few minutes, to see what would happen if we displayed an interest in the
product and to see if it would generate more ads. Two weeks later, we noted
another ad for an MBNA credit card. Financial institutions ads continued to
appear on both Ur and Bernadette's profiles occasionally. While it is clear
that Bernadette and Ur's gender profiles were cross-referenced with their different ages, we interpret the heavy presence of credit card companies in this
way. Both profiles were targeted by finance, credit card ads because these
companies use a much wider age criteria, targeting "the adult market," a very
large population. Ur and Bernadette did not attract many other advertisers and
so the empty space reserved for other advertisements became populated with

credit card ads. On Facebook, one either gets personalized ads or very general ads, depending on privacy settings, and according to our experiment, the data accumulated (or not) over time. Facebook's explanation for these ads is displayed online as a notification that can be checked. As we were informed: "One reason you're seeing this ad is that Capital One Canada wants to reach people who speak 'US' English" as well as people "18 years and older who live in or were recently in Quebec." The notification explains that this information was gathered from Facebook.

When we noted that ads were not appearing so readily in Ur's profile, we made more aggressive searches using Google and Yahoo: we would begin to buy a product, put it into our cart, and then close the page just before payment. Websites often use the "Facebook pixel"[4] (a piece of code) that allows them to target the people who have visited their website. On one occasion, we used "log in with Facebook" so the two sites would connect in some way. In our newsfeed, we also clicked on a product advertisement (a regular post) from Bathroom Basics from the group Age in Place. On Google, we clicked on ads that showed up related to our searches, for example, one brought us to Amazon for a search about "food ageing brain" (link to recipes books). Another search was about baby clothing and another was about life insurance for seniors. We targeted domains that seem like they invested heavily in marketing. However, none of this resulted in a plethora of ads in Ur's profile. Toward the end of the experiment, we started seeing sponsored posts (ads in the newsfeed; see figure 2.3). They seemed to be explained by the same

Figure 2.3   Bernadette, in the Ads Section. *Source:* Screen capture from Facebook by the researchers.

Your interests   Close

Business and industry   Travel, places and events   Family and relationships   Hobbies and activities   More ∨

Interests are determined based on your activity on Facebook, such as your engagement with certain Pages and ads.

Elderly care   Ageing   Credit cards   Capital One   Financial services   Aging in place

Cineplex   Healthways
Entertainment

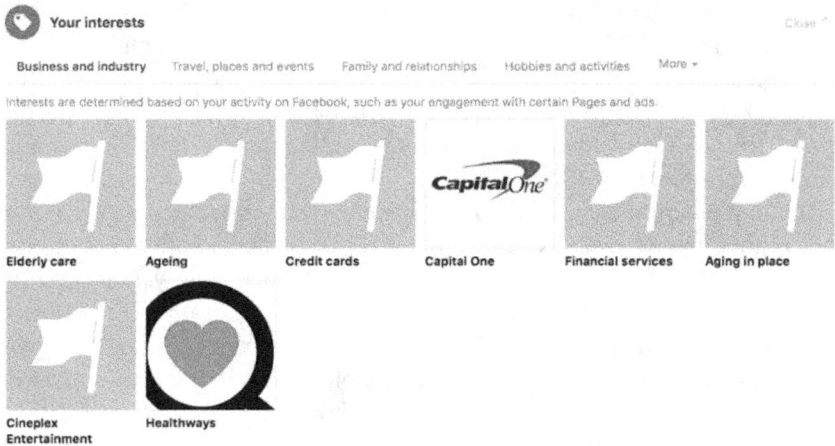

Figure 2.4   **Your Interests.** *Source:* Screen capture from Facebook by the researchers.

rationale behind the credit cards: companies aiming at a very large public. To date, Ur still receives sponsored content targeting Canadians who are not specific to those over the age of seventy but who belong to a very large and amorphous age group. For example, in December 2020, we saw a post advertising Metalbirds to 25-year-olds and over, across Canada.

In Bernadette's "Ads information" section in her profile settings, we saw *Interests* that made sense (e.g., "kitchen") and some items that did not, such as the name of an unknown, seemingly random, user and Cineplex Entertainment (see figure 2.4). Unfortunately, Bernadette's account was disabled. No specific reason was given, but we were directed to the help center stating the rules that were potentially violated (multiple accounts, fake name, fake person).[5]

To further investigate what these algorithms would take into account, we compared the results using one of the authors' actual personal profiles with Ur's profile. Maude clicked on a post from SEPAQ (national parks in Quebec) about "van life" (SEPAQ n. d.) then immediately got an ad from a company renting RVs a few minutes later. We then tried the same thing using Ur's profile. Ur liked and clicked on the link on SEPAQ's Facebook page about vanlife but never received an ad. It would seem that our behavior (clicks) did not provoke any algorithmic interest. And here it is worth noting that older adults, particularly those over seventy-two, are excluded more readily and are not a desirable targeted demographic for activities related in this instance to outdoor adventure. While Maude, thirty-five, could be imagined as interested in outdoor life, Ur, at age seventy-two, seems to be outside of this realm of potential interest to marketers. We noted the fact that Maude (who's been on Facebook for thirteen years or so) is also a fan of other groups related to hiking and had shown interest in SEPAQ posts before. Ur is also a fan of hiking (as shown in an article they shared), but for a much shorter period of time and hiking or nature did not figure in their interests (we checked the hobbies section of Your interests). As algorithms become refined, the behaviors taken into account including those related to age might become more precise.

To complement the experiment, we systematically documented the ads we were getting on our own profiles, taking screenshots from both our computers and phones. Maude and Kim both leave deep traces on the web as their privacy settings are not very restrictive and browsers, applications, and devices are allowed to collect a certain amount of data. For example, Kim, who uses applications on her phone, had tracking enabled, as well as location and her mic on, even though she does not use a profile picture.

Kim, the oldest member of the team, is not a regular user of Facebook, although she has an account. She has a Pinterest account, uses YouTube regularly to look up recipes, researches extreme camping, and she travels a great deal. She became interested in algorithms when she turned fifty-five and started to notice that she was being targeted with advertisements for bunions, despite no searches, purchases, or personal interest to reflect this. She has purchased clothing through Amazon, and regularly uses search engines (Chrome, Firefox, etc.). After traveling, she is often targeted with ads from airlines asking her to purchase an upgrade, clearly specific to her membership in her airline's point scheme and her use of their mobile app. Her interest in all things camping likewise produces a plethora of car ads. Most disturbing, however, are the ads that do not represent age but rather cultural projection of both age and gender-related concerns (such as belly fat), which regularly pop up on YouTube, alongside public service announcements on heart disease, diabetes, and arthritis, many of them connected to campaigns for specific over-the-counter medicines and products.

Scott, who is young and male, noted in this same period that he was targeted with advertisements for digital games, alcohol, cars, and travel. As a student in their twenties, many of the advertisements were reflective of cultural age expectations (such as drinking at parties) or inferred by the research he had been doing. Using these platforms across devices, Scott noticed that his recommended advertisements were influenced by other apps and device uses. For example, dating preferences curated specific content suggestions across platforms, and advertisements on his phone were clearly targeted toward his sexuality. Since Scott had curated his privacy settings higher on his personal computer compared to his phone, the information relay between his devices was exceedingly noticeable. Compared to the fictional profiles, the robust cross-device profile that Scott manages highlights the diverse interactions that algorithms play in relation to age and gender.

When one looks at the "why am I seeing this"[6] pop-up, we can learn who advertisers choose to target, for example using broad age categories.[7] Maude found (in her newsfeed) a Goodfood ad targeting people between 20 and 34 years old, speaking English and living in Quebec City, and a Gardasil (a vaccine against HPV) ad targeting women from twenty-seven to forty-four, for example. Upon checking the information Facebook provides, her "profile information" is presented as the reason she was targeted.

In summary, while cybernetic identities may be fluid, we can see here how age acts as a singular variable. In the case of the oldest old, our fictional character Ur, broad age categories seem to be used in advertising, especially when profiles, such as Ur and Bernadette's, are not very complex (having few to no friends, a small volume of activity). Furthermore, Ur, an older fictional character with no ascribed gender, received far less algorithmic attention than Bernadette, in her fifties coded as female. Of course, these fictional profiles have shallow lives compared to our "real" profiles. Indeed, it is likely that Bernadette was flagged when her profile was cross-referenced to an existing IP address. Ur, on the other hand, has a profile that in the context of Quebec—at this moment in time—means that they are more likely to not be of specific industries. Nevertheless, compared to gender, for example, there is a legal issue with age: some restrictions apply to advertisements, such as the targeting of children, for alcohol or credit card use. The age that is stated in one's profile is significant and performs in ways that intersect with other identity markers, such as gender, within the circuitry of algorithmic media.

LESSONS LEARNED

Lesson 1: Age Is Relational, Fluid—but Not Completely

This experiment indicates that online behavior, as measured by clicks and views, is only one aspect of how the tracking and imputation of age-related identity operates within algorithmic systems. What is most important are the prior and ongoing relations you have established with other users, as well as the history of one's movements through time. The less dynamic you are online, the less you will be targeted. Unlike our own "real" identities, Ur and Bernadette lack a robust social network (friends) that would have fed more data into the system (Cheney-Lippold).

When comparing the fictional profiles of Ur and Bernadette with our own, we indulge in a multitude of on and offline activities that they do not. We chat via Messenger because we have actual friends;[8] we have purchased offline and online for many years; we post photos of ourselves. Overall, we have left a far greater number of traces over the years. We also access our digital media via multiple devices/platforms that feed each other, as we noticed with Scott's mobile phone feeding profile information through the apps on it. Media, as an assemblage, takes on its full meaning when we look at the authors' profiles. Despite our attempts to gain some control over our online presence (through privacy settings for example), it seems impossible to not leave traces and fully stop the exchange of information between devices. This also translates into one of the difficulties we encountered during the experiment: how to start from a blank slate as a digital researcher.

Algorithms associate content to users based on content liked by similar users in the past (e.g., we saw this with the suggested groups Ur and Bernadette were getting), social filtering (which seemed much more relied on for the three authors than for the fictional profiles), and context (Tondello, Orji, and Nacke 2017). The small, but still significant, amount of related advertisements highlight the multifaceted parts of the algorithms, while also raising questions of our methodological process. As we engaged in the methodology, it led us to ponder Kitchin's suggestion that researchers engage in reverse engineering.

Because of its constant feedback loop, the cyber-identity of Ur and Bernadette can change following new information and new data (e.g., one's algorithmically assumed gender might change from male to female based on visiting certain websites). With varying statistical confidence, companies attribute a gender, race, age (etc.) to particular websites that are then used to infer the identity of its future visitors. However, the identity categories attributed to the site can also change according to the inferred gender of its visitors, based on an evolving collection of data (Cheney-Lippold 2011). Did Ur and Bernadette's *Your Interests* affect the identity the way that different algorithmic protocols read their identities? Were their fake ages and genders (unknown and female, respectively) believed by the system?

Algorithms *assess* actual behavior and then provide behavioral suggestions, and representations, based on what our culture as a whole thinks constitute age and ageing. Our two relatively inert and mostly dormant profiles attracted what the system understood as age-appropriate content: religion, life lessons, and extremely ageist cartoons and illustrations related to the chronological age we entered into the system. Our behavior did not indicate a previous interest in any of these activities, although as we noted the content of grandparent groups we liked often included posts with similar content.

John Cheney-Lippold (2011, 177) suggests that the "cybernetic categorization" of identity, which seems less fixed than non-digital categorization, "provides an elastic relationship to power" but still pushes people toward normalized behavior through the "constant redefinition of categories of identity," which he terms "soft biopower." He continues:

> If a certain set of categories ceases to effectively regulate, another set can quickly be reassigned to a user, providing a seemingly seamless experience

online that still exerts a force over who that user is. This force is not entirely benign but is instead something that tells us who we are, what we want, and who we should be. (Cheney-Lippold 2011, 177)

The soft biopower of algorithmic media is, for example, telling us what we should be interested in through advertising, through content suggestions on Amazon, Netflix, Spotify, and so on. On Facebook, it seems to be done in a pervasive way: suggested groups and ads appear in the same spot, whereas posts from groups we like, recommended posts, and sponsored posts all appear in the newsfeed with a very small legend, making this content almost seamless with the rest of the newsfeed. The conclusion we draw from our experience is that age matters in the world of algorithmic media: paradoxically, this new and powerful generator of cultural representations of age and ageing has a tendency to reproduce long-held stereotypes of old people or age as a series of shameful corporeal declines that must be targeted with a variety of anti-ageing products. As ageing subjects, we are "channeled" or web-lined into a universe of images and messages that reinforce what the algorithm thinks we are and who we should be, as we age.

Lesson 2: Interest-Driven Marketing, Personalization, and Age

Besides traditional criteria such as age, gender, and location, which is still relevant in Facebook advertising, there is a more personalized realm of marketing based on "interests." Ur and Bernadette's *Your Interests* section became filled with hobby categories, companies, and products, and for some we could not discern how they were related to their fake online presence. Likewise, as mentioned previously, Kim noted how her ads changed with her future desires (to build a rocket stove) as well as her past actual behaviors, including travel. The continual assessment of us, as users, by algorithms means that they react in real-time to our every move.

As noted previously, those who are fifty-five plus are often in a kind of "grey zone" (Sawchuk and Crow 2011) that does not differentiate between interests. However, there is also the fact that in Canada and Quebec, over 50 percent of those over seventy-two do not have Internet access, have the lowest usages of mobile phones, and have issues of access and affordability that is not just because they are incapable but are related to systemic factors such as the high cost of data in Canada. Is this why Ur was slighted, in a sense, by the system?

Returning to the question asked by the team of doctors on their assessment of the use of data on "older adults," this absence of older adults from the digital world also biases the system by eliminating from algorithmic purview significant numbers of people who are in their seventies. This was clearly revealed in the cases of Bernadette and Ur. As a fictional identity over the age of seventy, Ur may have been algorithmically irrelevant. Given their relative exclusion from the online world, people over seventy (Ur is seventy-two) may not be targeted as often as younger users through digital media. It is possible that this systemic absence means that there may not be enough data accumulated to train the algorithms associated with different social media platforms to create nuanced profiles of those who appear too old to matter. Again, the difference is striking between Ur's profile and ours. Longevity of use, multi-device use, and cross-platform actions all influence our algorithmic footprints. Connected with our ages and demographic interactions, these profiles become robust, personalized, and precise. Connecting the constructed profiles, which have very little data and few traceable interactions, with our own robust real profiles points to the temporal nature of algorithmic processes of data gathering and profiling. While age, as a variable, may function as a starting point for the algorithm, our fictional friends never gathered significant experiences online, over time. Ur and Bernadette may have been ascribed specific ages. They may have been assigned probable dates of birth. Yet, absent a history of activities and experiences, they never really aged.

Lesson 3: Methodological Implications

Throughout the experiment, we wondered how Facebook would respond to our rather passive engagements with its system. As Rob Kitchin (2017) states, proprietary systems have ways to know if "bots" are active and seeking to determine their algorithms. The same thing may be true for anyone creating a fictional profile. While living in cyberspace, Ur also was attached to an actual device. Ur's profile, for instance, was associated with Maude's computers. They were definitely linked to her phone. A couple of days after Bernadette's profile was set up, Facebook asked for a mobile phone number to send a code to verify our identity. After Maude gave her number, they were able to make connections between the two profiles. Shortly after, the account was banned from the platform because it was identified as fake. This raises questions about the methodological experiment that we set up, specifically around account upkeep and use. In a dual authentication environment, would you need to also purchase a burner phone to create a more authentically separate research profile?

What is clear is that we bring our algorithmic histories with us, as researchers. When we set up our fictional profiles, we fretted over what information Facebook could draw from us using our own computers. Somehow emptying the cache or downloading a new browser did not seem enough. As we all know from our experience as Internet users, companies like Facebook have ways to know if profiles are fake when you have no friends. What we discovered is that Ur and Bernadette were not targeted as much as expected, which reveals what gestures, activities, or features, such as clicks, purchases, for a network of friends, that have to already be there for a profile to seem interesting enough and authentic enough to target with nuance.[9] More often than not, when these fictional characters were targeted, they were included in a very large age cohort, such as "18-years-old and over." This suggests that the notion of using a "blank slate" to isolate algorithmic interpretations of the variable, "age" is close to impossible to implement, or would require extensive costs for analysis, or more active "fake" behavior which we did not choose to engage in. Understanding that our personal actions move between platforms, browsers, and technologies, researching algorithmic processes in digital spaces will always be impacted by the researchers' actions.

FAKING AGE AND ALGORITHMIC MEDIA

In conclusion, methodologically, it is evident that for reverse engineering experiments such as ours to be successful, they must be done over a long period of time to accumulate enough data. On a daily basis, real users will interact more (e.g., by using Messenger everyday with many different friends) than Bernadette and Ur ever did or ever would. Our personal profiles are the accumulations of a social media profile that not only targets us according to age but ages alongside and with us. At the same time, the busyness of our on and offline lives leads to a bombardment of numerous age- and gender-related ads daily. In contrast, Ur received few in two months. This is one limit of our experiment in reverse engineering using these fictional profiles.

And here we turn to a reflection on conducting research on algorithms and social media. As Kozinets (2015) writes, online access to vast amounts of archived social interactions changes the practice of ethnography. Into this vast and evolving ecosystem of social and individual data, of captured and emergent communications, netnography is positioned somewhere between the vast searchlights of big data analysis and the close readings of discourse analysis. Netnography entails the creation of specific sets of research positions and accompanying practices that are embedded in historical trajectories, webs of theoretical constructs, and networks of scholarship and citation. Yet what seems consistent, examined from the purview of age and ageing, is the deep ties of these new systems with existing systems and networks of cultural representations, including discourses that represent age and aging (Kozinets 2015).

Our experience in "faking age" underscores the relevance of combining more traditional methods like ethnography note-taking over long periods of time, embedding of the researcher in the settings, with new methods such as reverse engineering experiments to try and comprehend the black box of algorithmic media culture. The other limit of this method is that, as Seaver notes, "while reverse engineering can give some indication of the factors and conditions embedded into an algorithm, they generally cannot do so with any specificity" (quoted in Kitchin 2017, 23). We can, in other words, only *assume* the causes and conditions that have fed our observations on our own attempts to understand how and if we were targeted because of our differences in age. We know algorithms are working to make posts and ads for products, such as socks to prevent foot pain, show up. However, it is difficult to disentangle all of the data that is taken into account and how such data is prioritized precisely, by this system.

What this experiment in trying to "fake age" in algorithmic media reveals, most pointedly, is the limited social imaginary on age and ageing that is replicated in these systems—and how this is articulated to gender. Particularly troubling are the ways that the aging female body was targeted as a set of interlocking zones of shame, masked under a discourse of health and well-being. What is revealed, as well, is the exclusion of those over seventy from this system and their repercussions for the replication of ageist images. While ageing identities are associational and subject to the logic of target marketing that is based on a calculation of perceived interests, past representations of age continue to act as powerful determinants within the current algorithmic media assemblage.

NOTES

1. Despite the lack of focus on this area, there are some notable anecdotal examples such as in Cheney-Lippold (2017), who recounts the story of a friend of his: a female biologist in her twenties, her profiles classify her algorithmically as a man in her forties because of her viewing behaviors related to her occupation as a biologist.

2. For another critique of the role of devices like wearables, and gamification broadly, for implicating bodies in regimes of control, see Gómez, chapter 3, this book. — Ed.

3. Bucher (2012) gives the example of a woman who liked Taylor Swift and noticed that media inferred she was younger. In terms of deception, in both instances we did not engage in any online discussions. We created profiles to see what the algorithms would generate and interacted with the algorithm solely.

4. For more information about Facebook Pixel see https://www.facebook.com/business/learn/facebook-ads-pixel.

5. https://www.facebook.com/help/245058342280723.

6. This is achieved by clicking on the arrow at the top right corner of ads and sponsored posts on Facebook.

7. Although, as Nurik documents (chapter 1, this book) this might only tell part of the story, as these legends only reveal part of the targeting algorithm and only from

the *concrete* targeting values they enter. Additional *proxy* values (such as using postal codes to covertly target race) or *emergent* values (that the algorithm itself might attribute to a particular person) remain black-boxed. — Ed.

8. We couldn't friend random people for ethical reasons. Indeed, as mentioned, we minimized the behaviors of our two algorithmic avatars.

9. Some books and statistical reports recommend that digital marketers include seniors in their audiences (such as Stroud and Walker 2013; CEFRIO 2018a, b), but organizations do not seem to spontaneously use social media to target older people because although usage of digital media is increasing, a smaller percentage of them use these platforms and their various features (compared to younger cohorts). See https://cefrio.qc.ca/media/2188/netendances-2018-65-ans-et-plus_vf.pdf.

REFERENCES

Beer, David. 2017. "The Social Power of Algorithms." *Information, Communication & Society* 20, no. 1: 1–13. doi: 10.1080/1369118X.2016.1216147.

Bucher, Taina. 2016. "The Algorithmic Imaginary: Exploring the Ordinary Affects of Facebook Algorithms." *Information, Communication & Society* 20, no. 1: 30–44. http://www.tandfonline.com/doi/full/10.1080/1369118X.2016.1154086.

CEFRIO. 2018a. "Fiche Génération 2018." *NetTendances 2018, Vieillir à L'Ère Numérique.* https://cefrio.qc.ca/media/2188/netendances-2018-65-ans-et-plus_vf .pdf.

———. 2018b. "Quelle Relation les Aînés Entretiennent-ils avec le Numérique en 2018?" *NetTendances 2018, Vieillir à L'Ère Numérique.* https://cefrio.qc.ca/fr/e nquetes-et-donnees/netendances2018-vieillir-ere-numerique/.

Cheney-Lippold, John. 2017. *We Are Data: Algorithms and the Making of our Digital Selves.* NYU Press.

———. 2011. "A New Algorithmic Identity: Soft Biopolitics and the Modulation of Control." *Theory, Culture & Society* 28, no. 6: 164–181. http://journals.sagepub .com/doi/pdf/10.1177/0263276411424420.

Chun, Wendy Hui Kyong. 2019. "Queerying Homophily." In *Pattern Discrimination*, edited by Clemens Apprich, Wendy Hui Kyong Chun, Florian Cramer, and Hito Steyerl, 59–96. Minneapolis: University of Minnesota Press.

Diakopoulos, Nicholas. 2017. "Enabling Accountability of Algorithmic Media: Transparency as a Constructive and Critical Lens." *Studies in Big Data* 11. doi: 10.1007/978-3-319-54024-5_2.

Facebook. 2019. "Why Was My Account Deactivated?" *Facebook.* Accessed 2019. https://www.facebook.com/help/245058342280723.

Foucault, Michel. 1971. *L'ordre du Discours* [*Orders of Discourse*]. Gallimard.

———. 1976. *Histoire de la Sexualité* [*History of Sexuality*], vol. 1. Gallimard.

Gillespie, Tarleton. 2016. "Algorithm." In *Digital Keywords: A Vocabulary of Information Society and Culture*, edited by Ben Peters. Princeton, NJ: Princeton University Press. http://culturedigitally.org/2016/08/keyword-algorithm/.

Katz, Stephen, and Barbara L. Marshall. 2018. "Tracked and Fit: Fitbits, Brain Games and the Quantified Aging Body." *The Journal of Aging Studies* 45: 63–68. doi: 10.1016/j.jaging.2018.01.009.

Kitchin, Rob. 2017. "Thinking Critically About and Researching Algorithms." *Information, Communication and Society* 20, no. 1: 14–29. http://www.tandfonline.com/doi/full/10.1080/1369118X.2016.1154087#abstract.

Kozinets, Robert. 2015. *Netnography: Redefined*. London, UK: Sage Publications.

McKelvey, Fenwick. 2014. "Algorithmic Media Need Democratic Methods: Why Publics Matter." *Canadian Journal of Communication* 39: 597–613. https://www.cjc-online.ca/index.php/journal/article/view/2746/2495.

Noble, Safiya. 2016. *Algorithms of Oppression: Race, Gender and Power in the Digital Age*. New York: NYU Press.

Nye, Joseph S. 1990. *Bound to Lead: The Changing Nature of American Power*. Art of Mentoring Series (reprint ed.). New York: Basic Books.

Pasquale, Frank. 2016. *The Black Box Society: The Secret Algorithms That Control Money and Information*. Cambridge, MA: Harvard University Press.

Sandvig, Christian, Kevin Hamilton, Karrie Karahalios, and Cedric Langbort. 2016. "Automation, Algorithms, and Politics | When the Algorithm Itself is a Racist: Diagnosing Ethical Harm in the Basic Components of Software." *International Journal of Communication* 10. https://ijoc.org/index.php/ijoc/article/view/6182.

Sawchuk, Kim. 1996. "From Gloom to Boom: Age, Identity and Target Marketing." In *Images of Aging: Cultural Representations of Later Life*, edited by Mike Featherstone and Andy Wernick, 173–187. London: Routledge.

Sawchuk, Kim, and Barbara Crow. 2011. "Into the 'Grey Zone': Milieus that Matter." *Wi: Journal of Mobile Media* 5. http://wi.hexagram.ca/?p=69.

Seaver, Nick. 2013. "Knowing Algorithms." *Media in Transition 8*, Cambridge, MA. http://nickseaver.net/papers/seaverMiT8.pdf.

SEPAQ. n. d. "L'ABC de la Vanlife." *SEPAQ*. Accessed 2018. https://www.sepaq.com/blogue/abc-vanlife.dot?fbclid=IwAR3IdOSUBirp6hh5sCXeG_3kWz9SIo Beea-pkq98jfsipqC_M2WMYB6KJRM.

Striphas, Ted. 2015. "Algorithmic Culture." *European Journal of Cultural Studies* 18, no 4–5: 395–412. https://journals.sagepub.com/doi/pdf/10.1177/1367549415577392.

Stroud, Dick, and Kim Walker. 2013. *Marketing to the Ageing Consumer: The Secrets to Building an Age-Friendly Business*. London, UK: Palgrave Macmillan.

Taati, Babak, Shun Zhao, Ahmed B. Ashraf, Azin Asgarin, M. Erin Brown, Kenneth M. Prkachin, Alex Mihaildis, and Thomas Hadjistavropoulos. 2019. "Algorithmic Bias in Clinical Populations—Evaluating and Improving facial Analysis Technology in Older Adults with Dementia." *IEEE Access* 7: 25527–25534. doi: 10.1109/ACCESS.2019.2900022.

Tondello, Gustavo. F., Rita Orji, and Lennart E. Nacke. 2017. "Recommender Systems for Personalized Gamification." *Adjunct Publication of the 25th Conference on User Modeling, Adaptation and Personalization—UMAP '17*, 425–430. doi: 10.1145/3099023.3099114.

Chapter 3

It Was All Fun and Games

Gamewashing Automated Control

Sebastián Gómez

Think about all the points, badges, and leaderboards that populate your life. From the number of likes, views, or references you receive, to your hidden score on Tinder, your performance is evaluated through credit scores, insurance premiums, housing allocations. And if you are not good at this game, the algorithm will assign you a value as a bad risk assessment, shadowban you, or make you invisible. Yet, you accept being tracked and submit willingly to this competition because it is, apparently, the only way you can access the fun, enjoyment, or benefits of digitally articulated life. All these instances share the common characteristic of incorporating gameful experiences to entice and direct the actions of users, automating the trade of a reward for interacting with an algorithmic platform. Gamification—which this chapter considers the phylum composed by digital machine gambling, computer gaming, wargames/simulations, persuasive technologies, serious games, game-based learning, UI/UX design, content moderation, reward mechanics, and so on—is the use of game design elements in non-game contexts to drive engagement through the affordance of positive experiences (Deterding et al. 2011).

In the name of optimization and development, or simply in the name of making things a bit more fun, the notion of an inherently apolitical and innocuous tech (Hoofd 2007; Noble 2016; Abbing, Pierrot, and Snelting 2018) flows through the technological assemblages deployed to automate mechanisms surveilling, assessing, and directing human behavior. The problem emerges when this process helps perpetuate systems of oppression without a need for human oversight, since they become embedded in the technological procedures that automatically classify human populations and execute governmental policies. But with the excuse of making things better, technological research is often allowed to exclude already marginalized communities from almost every step of development and implementation.

Intersectional feminism has taken the lead in identifying how intersecting axes of oppression produce differentials in the lived experience of inequity that cannot be understood by separating subjectivities into single categories (Davis 2008). When confronted with digital automation, an intersectional approach helps describe how algorithmic biases are not a problem of, for instance, race and gender alone, but an issue stemming from Western philosophy and science being inserted in a concrete patriarchal, heteronormative, colonialist, and imperialist history that produces these axes of oppression and shapes the technology created within. In practice, intersectionality is a way of securing a situated knowledge, forcing the researcher to speak and be made accountable for their own positionality. As a method, it is more about how a problem is presented rather than offering a concrete theoretical framework. In order to do this, Mari Matsuda proposes "asking the other question":

> The way I try to understand the interconnection of all forms of subordination is through a method I call "ask the other question." When I see something that looks racist, I ask, "Where is the patriarchy in this?" When I see something that looks sexist, I ask, "Where is the heterosexism in this?" When I see something that looks homophobic, I ask, "Where are the class interests in this?" (quoted in Davis 2008, 70)

This chapter presents a model for positioning a research problem intersectionally when confronted with issues of automated control. Through the example of gamification as a computational logic embedded in the algorithmic architecture of techno-capitalism, I illustrate the connection of intersectional justice with algorithmic computation as an exclusionary practice, and computer gaming as a techno-culture disguising behavioral control through the principles of individual satisfaction, fun, and immersion.[1]

The first section introduces gamification as a case study that needs the attention of an intersectional critique. Gamification, through the automation of surveillance and feedback technologies experimented with since the inception of digital gaming, is capable of captivating users with its capacity to elicit positive feelings. The second section explores the concept of "fun" and proposes the term "gamewashing" to explain how the automation of fun experiences can still function as a mechanism of control and oppression, an aspect which has been dismissed in the application of gamified experiences. The final section offers an intersectional framing to explain how gamification has aided the automation of colonial, patriarchal, and heteronormative practices by demonstrating that even concepts like "fun" cannot be applied universally without first dismantling the power fantasies embedded in the notion of algorithmic control.

GAMIFYING BEHAVIORAL CONTROL

Gamification promises to transfer the engaging and persuasive power of games into a different context to encourage and direct the actions of employees, consumers, patients, students, and citizens (for a range of applications, see Walz and Deterding 2014). The idea is quite simple: You take design elements from known games and use them to make an otherwise dull activity fun. Though the process of turning drudgery into games is much older, the approach leaped into prominence around 2010 with the emergence of product-based approaches to gamification, with designers offering automated solutions for businesses promising to drive engagement ratings among consumers and employees. This approach is based on the Points–Badges–Leaderboards (PBLs) triad, illustrated on platforms such as Foursquare, LinkedIn, or Nike+. Extrinsic rewards, like accumulating points that can be traded for badges to show off to other users, are used to "hook" the user into climbing the leaderboards in a never-ending competition for first place. Businesses were presented with a low-cost software implementation that uses virtual rewards with exclusive value within the organization. The view of gamification as a simple add-on used to trigger predetermined responses, however, proved to be ineffective in the long run. It turns out people get used to—and annoyed by—discernible reward mechanics and constant feedback notifications relatively fast (Schüll 2014; Dubbels 2017). But growing beyond the corporate realm, research suggests that in the correct context, and through personalizing content along the interests of individuals, game design could prove useful in persuading people to change unwanted behaviors, learn better, experience simulated alternatives, and be prepared for potential futures (see Hamari, Koivisto, and Sarsa 2014; Bozkurt and Durak 2018). The recent "humanistic" (Deterding 2019) or "service-oriented" (Huotari and Hamari 2017) discourse in gamification tries to incorporate a more immanent approach to gaming, changing the angle toward user experience rather than specific design elements (Nicholson 2012). The idea behind this shift is to automate the creation of a personalized environment which allows for a gameful experience to emerge in concordance with the intrinsic interests of the user rather than depending on extrinsic rewards (Deterding 2015). Yet the low-cost and immediate benefits, even if short-lived, are sufficient to keep PBLs as the standard in gamification research (Bozkurt and Durak 2018; Koivisto and Hamari 2019). However, and regardless of the approach taken, the objective of gamification is still to influence human behavior (Bozkurt and Durak 2018), by eliciting positive experiences associated with games.

Gamification's behavioral management emerges as an unintended but welcomed evolution in digital gaming: the self-actualizing materiality of automated machines permits the design of an environment which creates the right conditions to activate an autonomous desire in users to *keep playing*. From the life-per-coin arcades to the digitalization of loyalty systems in casinos and credit cards, to the contemporary forms of subscription-based access, lootboxes, and microtransactions, the contribution of computer gaming to the algorithmic economy rests in its capacity to reward a user-generated action preemptively selected in concordance with the needs of the service provider. Gamification, in other words, seeks to automate behavioral control not only through the algorithmic configuration of digital technologies but also through the automation of the movements of human bodies. The expected behavior encouraged by gamified systems is a repetitive pattern of automated behavioral responses based on a psychosomatic need for enjoyment, resulting in the immersion into a system where "the lines between work and life, duties and pleasures, and company goals and personal goals are thereby increasingly blurred" (Mühlhoff and Schütz 2019, 238). Schüll, following Winnicott, describes "perfect contingency" as "a situation of complete alignment between a given action and the external response to that action, in which a distinction between the two collapses" (2014, 172). The overlapping of intrinsic and extrinsic goals is what has been popularly named, following Csikszentmihalyi (1990), as "flow." But as Schrank and Bolter describe: "Games or cultures that foster flow allow people to be perfectly subjugated within their systems. When a system is designed with optimal flow, people forget that they are being subjugated: their doubts and distractions are kept to a minimum, and all human labor is positively absorbed into the system" (quoted in Marcotte 2018). When the field of application of game design goes beyond the realm of actual games, what one encounters is a continuous subjectification *process*, an adaptive system that intersects algorithmic thinking and human subjectivity in an eco-technical setting aimed at collapsing together the goals of the system and the self.

Materially, gamification's strength is explained by two technological affordances developed in computer gaming which helped optimize the automation of human-data processing across computational machines. On the one hand, the inherent *surveillance requirements* (O'Donnell 2014) of computer gaming are extremely valuable considering how the deployment of more diverse and precise trackers needs their incorporation in daily routine to be not only voluntary but also wanted. On the other, the *feedback mechanisms* (Whitson 2014) experimented with in gaming proved to be an effective tool for creating and maintaining user engagement across digital platforms. But what makes gamification so effective is how surveillance and feedback work synchronously to drive engagement through an automated loop of emotional and affective manipulation. Algorithmic processes have the capacity to alter the environment in real-time in response to the player's behavior, causing feedback to materialize a new game-state that can subsequently inspire further actions (Leino 2003). The gamified platform is, in consequence, able to "modulate" the affective environment, increasing or decreasing rhythms and intensities to constantly and automatically mandate exclusive attention (Ash 2012). Simultaneously, the surveilled data is incorporated into the system, creating an optimal setting for the comparison of data points to the behavioral goals of the organization.

Automation allows these mechanisms to be replicated through algorithmic machines into every aspect of modern life. Not only is there a quantitative leap from previous analog forms of gaming in the sheer amount of individual information analyzed through big data processes, but also in what can be qualitatively done to it through those analytics. The blackboxed processes of recursive and ever-more-complex algorithms assign values to inputs and categorize according to their own logic, giving back to the user only what they need to know to continue playing. Meanwhile, a record of those actions becomes available to those with different levels of access to the platform. In the end, digital platforms are capable of modulating feedback with appropriate rewards in direct response to the information given by the users, using game elements in a continuous loop that secures ever-increasing engagement.

Online marketing, management, governance, and user experience design, among others, are beneficiaries of the persuasive technologies honed within gaming. Despite not having explicitly stated the use of gamification strategies, social media giants such as Facebook, Twitter, and Google are no doubt aware of the profit which can be obtained from applying game elements to their platforms. If not always a clear example of gamification, social media's role in manipulating behavior at mass scales belongs to the same genealogy of persuasive technologies originating within digital games. Deibert (2019) describes, for instance, how the outcomes of major political events[2] have been influenced by social media in their effort to monopolize user attention by using elements of game design:

> As players [play the game repeatedly], slowly becoming addicted, the game's application learns more and more about their devices, interests, movements, and so on. Social-media platforms even sense when you are disengaged and have designed techniques and tools to draw you back in: little red dots on app icons, banner notifications, the sound of a bell, a vibration. (29–30)

With the focus on driving engagement to their platforms, social media worries first and foremost to motivate people to generate and share content, regardless of the message. Users are encouraged to share data in exchange for a forgettable yet addictive dopamine boost, thus associating the action of participating with the fun of a gameful experience.

Naturally, gamification's goal of persuasion has been heavily criticized by different academic sectors as manipulative (O'Donnell 2014); alienating (Fuchs 2014); exploitative (Ferrer-Conill 2018); unethical (Perrotta et al. 2019); or, as Ian Bogost carefully phrased it, "bullshit" (2014). Yet, a preemptive dismissal of gamification research by critical studies has caused a lack of attention toward its impact on social justice. The situation is worsened by an attitude of indifference coming from game designers and theorists alike, so far as game studies have distanced themselves from gamification work, which is seen as an oversimplification of serious game research (O'Donnell 2014). And although the application of an intersectional framework in game studies is being valiantly put forward by scholars and designers from the perspective of feminist studies, queer studies, postcolonial studies, disability studies, and other fields concerned with social justice (e.g., Ruberg and Phillips 2018; Mukherjee and Hammar 2018), the focus has remained confined to studying specific cases of intersectional oppression in individual video games and demographics. However, the governmental processes strengthened by gamified algorithms escape already constituted individuals, as the society of control (Deleuze 1992) atomizes and scatters elements of power differentials to form a more tightened immersive economy. When the automation of game elements meets everyday life, a critical response must take an intersectional approach, as the experience of everyday life differs widely among different populations, with the historically subordinated being particularly vulnerable to any intensification in systems of behavioral influence and control.

Gamification, despite its fair share of criticism, has defended itself by declaring to mean well. Benefits to engagement in education, self-management, or health care are repeatedly cited as examples of applications where any accusation of exploitation or manipulation is put to rest when the focus is on well-being, improvement, and enjoyment. If everybody is having a good time voluntarily, and share the goals declared by the gamification provider, what could be the problem? But as the phrase goes: it is all fun and games until someone loses an eye.

IN THE NAME OF FUN

So far, I have presented gamification as a case study in need of an intersectional reading. The way in which automated technologies have incorporated gameful experiences to encourage users to interact willingly and compulsively with them presents a problem when the potential negative effects are sidelined in lieu of the benefits they deliver to the algorithmic economy. In this section, I provide examples of the consequences of perpetuating an obsession with behavioral control, even when done in the name of fun.

Fun and the enjoyment which accompanies its emergence is axiomatic in the formulation of games research (Ruberg 2015). It is also the argument gamification proponents use to differentiate it from policing and manipulative applications of other automated technologies. For instance, Deterding explains:

> What sets games apart from other environments is that they are deliberately designed to afford positive experiences. Hence, what gamification can bring to management is a particular design practice: re-organizing processes, products, and services to afford positive, well-being-supporting experiences for all stakeholders to drive organizational goals. (Deterding 2019, 3)

Also,

> [g]iven that game design is the practice of creating enjoyable interactions, it stands to reason that it holds something of interest to any domain where interaction is designed and the goal is to make it more enjoyable. (Walz and Deterding 2015, 9)

And so, the public acceptance of social experiments through behavioral interventions is weighed against the benefits of accessing the technological advantages needed to obtain access to these positive experiences. O'Donnell, in an ethnographical note, comments:

> Games leverage a variety of data sources as input to make these interactions meaningful. In a way, games cannot help but surveil the user: it is how a game reacts to its player. [. . .] Games were already surveillance systems and the expansion of that surveillance in the name of "fun," seemed a small trade-off for players. The opacity of these systems was no different than the opacity of a game's underlying rules and systems. It was normal. (O'Donnell 2014, 350)

The focus on positive experiences linked to the fun associated with play throws a veil over "governance projects that masquerade under the rhetoric of being just a fun game" (Whitson 2014, 354). The automation of surveillance and feedback, the key technological affordances of digital gaming, goes under the radar of critical readings so long as not even the ones negatively affected complain about it, since it is not easy to see how much control it has over our actions if there is a strong positive feeling associated with its implementation.

I propose the concept "gamewashing" to explain this phenomenon. Gamewashing, a term casually and sporadically used as a criticism to badly implemented gamification solutions, describes superficial game mechanics that are carelessly employed to mask an otherwise alienating activity with fun, without altering any aspect of the original system. But as game elements populate everyday life, the term "gamewashing" needs to expand as a conceptual tool to describe instances of automated control which are uncritically accepted not only because they are masked as something fun but because the fun aspect increases their effectiveness. At the same time, while the center of attention remains on fun and "positive experiences," the negative outcomes of gamification and the general "ludification of society" (Raessens 2006) remain largely unexplored (see Koivisto and Hamari 2019, 205), as "negative feelings" are dismissed "by reframing them as stepping stones on the road to success" (Ruberg 2015, 109). In the name of fun, enjoyment masks the implementation of systems of governance which do not need to directly discipline subjects, but simply control behavioral responses by designing a positive, immersive atmosphere. Talking about the immersive economy, Mühlhoff and Schütz explain that "such measures exist to produce suitable affective dispositions in co-workers, which allows for a form of corporate governance that operates almost exclusively by 'positive' forces" (Mühlhoff and Schütz 2019, 239). These positive forces, despite *feeling* like fun, can automate the strategies of control examined in the previous section.

Even formulating the possibility of a negative experience in relation to gaming is met with a strong resistance, both from researchers in game studies, who seem reticent to condemn negative feelings in game cultures for fear of re-igniting the "moral panic" over video game violence (Evans and Janish 2015), and from the neoconservative sectors within digital cultures. The latter, however, has demonstrated the power gameful experiences have in encouraging people to mobilize toward a goal, while causing damage in the process. The Gamergate movement (Mortensen 2016), which marked the declaration of the "culture wars" on the Net and the radicalization of white men against what they perceive as a threat to the "apolitical" experience of their "just-for-fun" games, is a strong example of how far gaming cultures will go to defend the medium they love from any criticism. During the Gamergate event in 2014, conspiracy theories spread on forums like 4chan, starting a belief in a secret coalition between media outlets to push for a "liberal agenda" within game development and infiltrate the medium white men thought to be their exclusive turf. Women were specifically targeted in response to the Gamergate claims that female journalists and designers were attempting to change games using their sexual power to confuse and mislead (straight male) developers. The rampant misogyny masqueraded under a call to defend media integrity, which led to instances of harassment and death threats to journalists, designers, and scholars who dared criticize games.

However, the protection of gameful experiences in the name of fun is not the only aspect of oppression that Gamergate exemplified. It is also a matter of algorithmic justice, so far as the mobilization of neoconservative, alt-right, and fascist ideals has been automated by affective loops in social media. Eco-chambers amplify extreme positions based on raw feelings of fear and insecurity and transform them into positive experiences by rewarding active participation. As game elements automate this pattern, ludic experiences emerge in the narratives shared on platforms like 4chan and Reddit. The organization of these movements acquires the structure of a game: It starts with a call to arms to complete an epic quest, making participants compete for the likes, shares, and views measuring the individual contribution to the cause, and eventually winning by "triggering," silencing, or conquering the perceived enemy. The conspiracies run wild, up to the point where, for example, Gamergate "claimed Jews and western academics have joined forces to pacify White men, and planned to hand the power of the 'western world' to the Jews or Islam by encouraging politically correct digital games" (Mortensen 2016, 788). Since then, the model for mobilization has been repeated on numerous occasions. In 2020, the "culture war" has taken an unprecedented global scale with conspiracy theories like QAnon—appearing within the same crowd in 4chan—assuming the shape of an Alternate Reality Game (Flam 2021), with millions of people participating in a hunt for clues to uncover the "truth" behind the COVID-19 pandemic, 5G networks, vaccines, and elites consuming the blood of children. Eventually, the combination of conspiracy theories and game elements driving fanatical engagement led to a resistance around the world to public health policies, thus hindering the efforts to slow the spread of the virus while seeding disbelief in science, government, and the common good. In a grim turn of events, Gamergate and QAnon show that the hopes of play transcending the realm of videogames and uniting people under a common goal to fix a "broken reality" (McGonigal 2011) became true, although not exactly for the purpose of the betterment of humanity that McGonigal had envisioned. Nonetheless, the gamification ideals which emerged from McGonigal's sentiment proved to be possible and immensely effective. It is feasible to transfer game elements into wider non-gaming contexts, recreating the experience of fun in a fully immersive environment, motivating people to act. Yet, the exclusive focus on fun hid in plain sight the possibilities of an algorithmic amplification of harm.

From an intersectional perspective, the interest in mobilizing for a cause like Gamergate reveals that what is considered positive and fun in the algorithmic fantasy of control inherited by digital games is related to the practices that have been fostered by a colonial, patriarchal, and cis-heteronormative system. According to Mitchel, digital games embody the implicit political assumptions in technology enabled by an algorithmic power which believes "the world can and should be controlled, both because it *is* controllable and because the lack of control poses a *threat*" (Mitchell 2018, 51; italics in original). Through the expansion of gameful experiences to all aspects of everyday life, the power fantasy enabled by algorithmic control infects the subjective experiences particular to the needs of automation, invoking the perpetuation of the colonial logic of "exploration, expansion, exploitation, and extermination" enforced by game procedures (Ford 2016). As a result, "gamers are set up to be colonial forces. [Playing video games is] about individuality, conquering, and solving. Feeling empowered and free at the expense of the world" (Brice, quoted in Marcotte 2018, n. p.). Game elements double as elements of control, feeding the call to action from fear and rewarding with positive experiences those who follow that call, automating a pattern of intensification in the defense of systems of privilege.

These examples show how gameful experiences are capable of escaping the confinements of hardware to motivate and organize people toward the goal of keeping the systemic privilege they had secured, causing harm to the perceived enemy while having fun doing it. Rather than a contradiction, gamewashing oppression is at the core of the contemporary experience of fun. Unfortunately, the veil cast by gamewashing often prevents the proponents of including gamefulness in everyday life to see that the positive feelings associated with play in our historical context belong to a game lineage obsessed with control by conquest and dominance.

FUN IS NOT UNIVERSAL

As a final exploration of the gamification of society, I problematize the concept of fun by questioning how it is applied as a universal category. The focus on fun presented so far assumes that everyone reacts in a similar way to positive stimuli, neglecting the historical systems of oppression that differentiate the everyday experience of communities and individuals. The importance of "asking the other question" becomes clear, especially when those who are affected by negative outcomes are not considered in the planning or participation of these gamified processes.

Gamewashing affects the formulation and interpretation of gamified applications, as the exclusive focus on positive experiences creates a self-fulfilling prophecy that naturalizes the behavioral traits rewarded by the power fantasies enabled and measured by those same experiments. "So despite (or perhaps because of) the good intentions of game designers and publishers," explains Hoofd (2007, 13), "these [serious] games then, in fact, exhibit the doubling of the colonialist logic that inspired humanist narratives of progress."

For instance, in Bozkurt and Durak's (2018) table surveying the main approaches in gamification research to explain *if* and *why* gamification is effective, we see that the most used theories belong to behavioral psychology and, in particular, self-determination theory and flow. These methodologies consider what materialist psychologists call a "within-subject" approach, which means they do not deal explicitly with processes of subjectification and power relationships co-constituting the already territorialized forms of human psychology (see Søndergaard 2002), or in other words, they are used ignoring a context where "the problems of execution are historically situated and entangled with the contingent forces of machines, bodies, institutions, military labour practices and geopolitics, rather than simply a set of instructions that are outside of life" (Pritchard, Snodgrass, and Tyzlik-Carve 2018, 11). When dealing with automated technologies created under the premise of control permeating algorithmic thinking, these theories dismiss the complex reality of human behavior in relation to computing practices, while game-washing behavioral interventions provide an illusion of scientific objectivity to their implementation.

Let me illustrate with some examples. *Gamification 101* (Bunchball 2012) is a handbook distributed for free by the gamification consultancy agency Bunchball, which describes a direct causation between "basic human desires" (2012, 6) and specific game elements. It claims, for instance, that a universal instinct to compete can be satisfied through the implementation of leaderboards, and therefore justifies its inclusion in gamification design. This kind of rhetoric is not isolated, since the pressure to classify behavioral motivations of different "player types" for marketing purposes has led researchers to identify the desire to play in a human need for "achievement, exploration, sociability, domination and immersion" (Hamari and Tuunanen 2014, 46). In other words, it is implied that satisfying the player's natural need to dominate *causes* a positive experience. But in doing so, gamification ignores the historical context where exploration, competition, success, domination, and conquest have been selected as desirable traits and embedded in subjectification processes that form individuals who will later enjoy and be rewarded for exhibiting those traits. Through this approach to behavioral traits, for example, the governmentality commanding individuals to succeed is naturalized as inherent to the human condition rather than a consequence of social pressures specific to a neoliberal context. I contest, then, that leaderboards do not satisfy an inherent need to compete but rather reward an artificial command to dominate and control, and thus amplify it as the body associates it with a positive experience. The anthropocentric, heteropatriarchal, colonial, and normative logics embedded in algorithmic platforms automate a reward system enhancing the behavioral traits privileged by centuries of an universalist and interventionist "Humanism" (Braidotti 2016).

With an algorithmic architecture automating rewards to actions of conquest and dominance, human well-being is automated by computational systems through a pathologized promise of self-fulfillment *if and only if* the competition is eliminated and any personal fault or deviance has been erased. One could even consider QAnon's hero, Donald Trump, as the model for the ultimate "gamer," with his relentless need for winning at all cost, and his casting of vulnerabilities, including to diseases like COVID-19, as weaknesses to be dismissed. The automation of gameful experiences, emerging exclusively from the positive feeling of a dopamine hit, neglects the damage an addiction to control can cause. In this sense, the need for conquest and dominance satisfied by game design is already embedded in the algorithmic logic of control through its inherent need to surveil and reflect real-time feedback to the user. The libidinal loop of data-giving and assessment-receiving in an algorithmic ecosystem, repeated in micro-intervals and maintained as a consistent, objective collection of real-life data, is enough to force the production of value to move toward automated systems of competitive gaming which transform everyday life into an actual survival of the fittest. "Fit," moreover, is determined by the black-boxed algorithmic processes and uncriticized biases in technological research and development. This is because Western research is modeled on the ideal embodied in the figure of a white, cis, straight, able-bodied, and wealthy man (Braidotti 2013; Davis 2008), each axis acting independently in setting the standard for what is considered the norm, amplifying each other as they intersect. And when these mechanisms materially control access and visibility in the name of individual well-being, once more are the most vulnerable the ones to first be identified, classified as a group, excluded, and potentially exterminated. When human identity resists the subjectification forces of a humanist taxonomy inherited by normative computation, non-normative bodies and behaviors become problematic when confronted with algorithms trained to accept social reality as an objective and apolitical dataset (Abbing, Pierrot, and Snelting 2018). The affective atmospheres generated by the gameful environment are co-opted by existing power differentials between bodies, creating preemptive scars of oppression through the forced identification of individuals to preexistent, hierarchized identity groups. The historical disadvantages over wealth, race, ethnicity, sexuality, sex, gender identity, gender expression, physical and neurological ability are programmed as objective categories for algorithmic execution.

In the case of automation, to ask the other question means to dismantle the idea that digital technologies are objective tools of identification which can be applied uncritically to the entirety of human experience, since there is always an excluded other when a universal application is assumed. An intersectional, materialist feminism is no longer rooted in the identity of fixed individuals, but in the subjective experience of living bodies (Staunæs 2003), offering a chance to challenge the processes of normative domination at the source of struggles for social justice. Non-identities, troubled identities, queer, and non-binary identities have challenged the supremacy of belonging and highlighted the plasticity of subjective experience (Ahmed 2006) while confronting the relativism of postmodern representationalism through a deep-rooted belief in the material embodiment of human experience (Braidotti 2011).[3] Asking the other question allows us to shine a light on the biases and exclusions embedded in automated technologies, which are never innocuous or apolitical, even when done in the name of fun.

CONCLUSION: ON THE NEED TO EMANCIPATE GAMING

The gamewashing effect of the gamification process establishes mechanisms which justify behavioral interventions with the excuse of enabling fun. They are sometimes; but as the field largely run by the same privileged demos naturally missed, they are conceived and perceived as fun by those who do not risk anything when playing. Digital gaming for us at the margins is a painful, risky experience of self-management, forced misidentification, and negation of minoritarian subjectivities. And yet we want to play anyway. Because the desire to play games goes beyond the specific configurations it acquires through the organizing forces of humanist, colonial normalization. If digital connections are there to create an ever-changing environment, we will adapt as desiring-machines do to survive. And there is the true emancipatory power of gaming.

How do we create gameful experiences in a form other than executive domination? Can we simply stop using such systems and start alternatives? Naturally. Queering, decolonizing, and intersectional approaches to gaming have already started such projects (e.g., Ruberg 2019, 2015; Marcotte 2018; Freedman 2018; Murray 2018). For instance, Ruberg (2015) proposes a form of queer gaming that focuses on non-normative play through the deliberate choice of not having fun, while Khaled (2018) is experimenting with "reflective game design" by breaking patterns of immersion and highlighting the critical potential of gaming. The disruptive power of countergaming (Galloway 2006), embodied in the projects of alternatives to the gamifying processes, is an example of rethinking not only the cultures and the representation of non-normative subjectivities but how to create games and computation differently (Freedman 2018). Queer theory and intersectional feminism have provided key understandings on the concept of troubled and non-identities: queer as in a "semantic openness and expressive possibility that can challenge (or productively deconstruct) normative realism" (Freedman 2018, n. p.). Confusing, dynamic, outside the binary, and beyond the oft-linear spectrum of identity politics, this position resists the assigned identification of gaming systems, creating alternative modes of play, such as disruptive play, transgressive play, emergent play, and other troubled gameplays. But accepting the margins is something that is undoubtedly marked by the failure to develop alternatively or sustainably to resist the neocolonial domination that ultimately has put the entire planet in crisis. If we, as critical gamers, relegate ourselves to constructing alternatives over a rotten structure of imperial gaming—at the margins of the industry, of academia, of the target audiences—we risk letting normative and governmental strategies run wild with a technology we know can change lives so profoundly.

Gamification has been ignored as industry nonsense and let go unchallenged into governmental projects and tech monopolies swiping entire elections, democracies, and populations with the pretext of gamified platforms simply fulfilling an inherent need for fun. Creating an alternative, as intersectionality demonstrates, does not prevent the consequences of an uncritical acceptance of patriarchal, heteronormative, and colonialist logics in gamification at different scales of social organization. Computer gaming needs to be redefined by emancipating the concept of gaming and playing from the gamewashed vocabulary of a humanist tradition which naturalizes games as a diagram for identities generated in conflict. Beyond playing differently, it is important to liberate play: "To the extent that play is creative and emancipatory, it is of critical importance to be able to dislodge play from normative structures, especially those that arise within gaming" (Knutson 2018, np.). If identities, scores, rules, and goals continue to be assigned rather than self-determined, playing with machines will keep being shaped by the "games of Empire" (Dyer-Witheford and de Peuter 2009).

Algorithms, as well as identities, ought to be problematized and never taken as fixed, agentless, servant tools for an external mastermind. Algorithms, in their capacity to calculate potentialities and execute realities, cannot help but infect computation with traces of non-normative gaming. The randomness emerging from their processes (Parisi 2015) is a glimpse of a form of gaming not constrained by the heteropatriarchal and colonial logic of fun and immersion. The seeds are there, we just need to take games seriously.

NOTES

1. For further discussions of algorithmic exclusion on platforms such as Facebook, see Sawchuk, DeJong, and Gauthier, chapter 2, this volume for exclusion related to age, and Nurik, chapter 1 this book, for exclusion related to protected status groups in the United States — Ed.

2. When the content shared in these platforms involves misinformation, it is possible to understand how in 2020 conspiracy theories spread so widely and with such a passionate response during the COVID-19 pandemic.

3. For discussions of artistic works that challenge these material embodiments from posthumanist and queer perspectives, see Bouma, chapter 10; as well as van Veen, chapter 12, this book, with the last also in a gaming context. — Ed.

REFERENCES

Abbing, Roel Roscam, Peggy Pierrot, and Femke Snelting. 2018. "Modifying the Universal." In *DATA Browser 06: EXECUTING PRACTICES*, edited by Helen Pritchard, Eric Snodgrass, and Magda Tyzlik-Carve. Open Humanities Press. data -browser.net/db06.html.

Ahmed, Sara. 2006. "ORIENTATIONS: Toward a Queer Phenomenology." *GLQ: A Journal of Lesbian and Gay Studies* 12 (4): 543–74. https://doi.org/10.1215/1 0642684-2006-002.

Ash, James. 2012. "Attention, Videogames and the Retentional Economies of Affective Amplification." *Theory, Culture & Society* 29 (6): 3–26. https://doi.org /10.1177/0263276412438595.

Bogost, Ian. 2014. "Why Gamification Is Bullshit." In *The Gameful World: Approaches, Issues, Applications*, edited by Steffen P. Walz and Sebastian Deterding, 65–80. Cambridge, MA: MIT Press.

Bozkurt, Aras, and Gürhan Durak. 2018. "A Systematic Review of Gamification Research." *International Journal of Game-Based Learning* 8 (3): 15–33. https:// doi.org/10.4018/IJGBL.2018070102.

Braidotti, Rosi. 2011. *Nomadic Theory*. New York: Columbia University Press.

———. 2013. *The Posthuman*. Cambridge: Polity.

———. 2016. "Posthuman Critical Theory." In *Critical Posthumanism and Planetary Futures*, edited by Debashish Banerji and Makarand R. Paranjape, 13–32. India: Springer. https://doi.org/10.1007/978-81-322-3637-5.

Bunchball. 2012. "Gamification 101: An Introduction to Game Dynamics." https:// www.bunchball.com/gamification101.

Csikszentmihalyi, Mihaly. 1990. *Flow: The Psychology of Optimal Experience*. New York, NY: Harper and Row. https://doi.org/10.2307/1511458.

Davis, Kathy. 2008. "Intersectionality as Buzzword: A Sociology of Science Perspective on What Makes a Feminist Theory Successful." *Feminist Theory* 9 (1): 67–85. https://doi.org/10.1177/1464700108086364.

Deibert, Ronald J. 2019. "The Road to Digital Unfreedom: Three Painful Truths About Social Media." *Journal of Democracy* 30 (1): 25–39. https://doi.org/10.1353 /jod.2019.0002.

Deleuze, Gilles. 1992. "Postscript on the Societies of Control." *October* 59 (1): 3–7. https://doi.org/10.2307/778828.

Deterding, Sebastian. 2015. "The Lens of Intrinsic Skill Atoms: A Method for Gameful Design." *Human-Computer Interaction* 30 (3–4): 294–335. https://doi.org /10.1080/07370024.2014.993471.

———. 2019. "Gamification in Management: Between Choice Architecture and Humanistic Design." *Journal of Management Inquiry* 28 (2): 131–36. https://doi .org/10.1177/1056492618790912.

Deterding, Sebastian, Kenton O'Hara, Miguel Sicart, Dan Dixon, and Lennart Nacke. 2011. "Gamification: Using Game Design Elements in Non-Gaming Contexts." *Conference on Human Factors in Computing Systems—Proceedings*, 2425–28. https://doi.org/10.1145/1979742.1979575.

Dubbels, Brock Randall. 2017. "Gamification Transformed." In *Transforming Gaming and Computer Simulation Technologies across Industries*, edited by Brock Randall Dubbels, 17–47. Hershey, PA: IGI Global. https://doi.org/10.4018/978-1-5225-1817-4.ch002.

Dyer-Witheford, Nick, and Greig de Peuter. 2009. *Games of Empire. Global Capitalism and Video Games*. Minneapolis, MN: University of Minnesota Press.

Evans, Sarah Beth, and Elyse Janish. 2015. "#INeedDiverseGames: How the Queer Backlash to GamerGate Enables Nonbinary Coalition." *QED: A Journal in GLBTQ Worldmaking* 2 (2): 125. https://doi.org/10.14321/qed.2.2.0125.

Ferrer-Conill, Raul. 2018. "Playbour and the Gamification of Work: Empowerment, Exploitation and Fun as Labour Dynamics." In *Technologies of Labour and the Politics of Contradiction*, edited by Paško Bilić, Jaka Primorac, and Bjarki Valtýsson, 193–210. Cham: Springer International Publishing. https://doi.org/10 .1007/978-3-319-76279-1_11.

Flam, Faye. 2021. "How the QAnon Conspiracy Seduces Normal People." *Bloomberg*, January 2021. https://www.bloomberg.com/opinion/articles/2021-01-30/the -science-behind-qanon-how-conspiracy-can-seduce-normal-people.

Ford, Dom. 2016. "'EXplore, EXpand, EXploit, EXterminate': Affective Writing of Postcolonial History and Education in Civilization V." *Game Studies* 16 (2). http:// gamestudies.org/1602/articles/ford.

Freedman, Eric. 2018. "Engineering Queerness in the Game Development Pipeline." *Game Studies* 18 (3). gamestudies.org/1803/articles/ericfreedman.

Fuchs, Mathias. 2014. "Gamification as Twenty-First-Century Ideology." *Journal of Gaming & Virtual Worlds* 6 (2): 143–57. https://doi.org/10.1386/jgvw.6.2.143_1.

Galloway, Alexander R. 2006. *Gaming: Essays on Algorithmic Culture*. Minneapolis, MN: University of Minnesota Press.

Hamari, Juho, Jonna Koivisto, and Harri Sarsa. 2014. "Does Gamification Work?—A Literature Review of Empirical Studies on Gamification." *Proceedings of the Annual Hawaii International Conference on System Sciences*, 3025–34. https://doi .org/10.1109/HICSS.2014.377.

Hamari, Juho, and Janne Tuunanen. 2014. "Player Types : A Meta-Synthesis." *Selected Articles from the DiGRA Nording 2012 Conference: Local and Global-Games in Culture and Society*, 29–53. https://doi.org/10.1111/j.1083-6101.2006.00301.x.

Hoofd, Ingrid M. 2007. "The Neoliberal Consolidation of Play and Speed: Ethical Issues in Serious Gaming." *Critical Literacy: Theories and Practices* 1 (2): 6–15.

Huotari, Kai, and Juho Hamari. 2017. "A Definition for Gamification: Anchoring Gamification in the Service Marketing Literature." *Electronic Markets* 27: 21–31. https://doi.org/10.1007/s12525-015-0212-z.

Khaled, Rilla. 2018. "Questions Over Answers: Reflective Game Design." In *Playful Disruption of Digital Media*, 3–27. Singapore: Springer. https://doi.org/10.1007/978-981-10-1891-6_1.

Knutson, Matt. 2018. "Backtrack, Pause, Rewind, Reset: Queering Chrononormativity in Gaming." *Game Studies* 18 (3): 1–15.

Koivisto, Jonna, and Juho Hamari. 2019. "The Rise of Motivational Information Systems: A Review of Gamification Research." *International Journal of Information Management* 45 (June 2017): 191–210. https://doi.org/10.1016/j.ijinfomgt.2018.10.013.

Leino, Olli Tapio. 2003. "Death Loop as a Feature." *Game Studies* 12 (2). http://gamestudies.org/1202/articles/death_loop_as_a_feature.

Marcotte, Jess. 2018. "Queering Control(Lers) Through Reflective Game Design." *Game Studies* 18 (3): 1–20. gamestudies.org/1803/articles/marcotte.

McGonigal, Jane. 2011. *Reality Is Broken: Why Games Makes Us Better and How They Can Change the World*. New York, NY: The Penguin Press.

Mitchell, Liam. 2018. *Ludopolitics: Videogames against Control*. Wichester: Zero Books.

Mortensen, T. E. 2016. "Anger, Fear, and Games: The Long Event of #GamerGate." *Games and Culture* 13 (8): 787–806. https://doi.org/10.1177/1555412016640408.

Mühlhoff, Rainer, and Theresa Schütz. 2019. "Immersion, Immersive Power." In *Affective Societies*, 231–40. Routledge. https://doi.org/10.4324/9781351039260-20.

Mukherjee, Souvik, and Emil Lundedal Hammar. 2018. "Introduction to the Special Issue on Postcolonial Perspectives in Game Studies." *Open Library of Humanities* 4 (2): 1–24. https://doi.org/10.16995/olh.309.

Murray, Soraya. 2018. "The Work of Postcolonial Game Studies in the Play of Culture." *Open Library of Humanities* 4 (1): 1–25. https://doi.org/10.16995/olh.285.

Nicholson, Scott. 2012. "A User-Centered Theoretical Framework for Meaningful Gamification." *Games+ Learning+Society*, 1–7. https://doi.org/10.1007/978-3-319-10208-5_1.

Noble, Safiya Umoja. 2016. "A Future for Intersectional Black Feminist Technology Studies." *The Scholar & Feminist Online* 13 (3). sfonline.barnard.edu/traversing-technologies/safiya-umoja-noble-a-future-for-intersectional-black-feminist-technology-studies/.

O'Donnell, Casey. 2014. "Getting Played: Gamification, Bullshit, and the Rise of Algorithmic Surveillance." *Surveillance and Society* 12 (3): 349–59.

Parisi, Luciana. 2015. "Instrumental Reason, Algorithmic Capitalism, and the Incomputable." In *Alleys of Your Mind: Augmented Intelligence and Its Traumas*, edited by Matteo Pasquinelli, 125–37. Mitton Keynes: Meson Press.

Perrotta, Carlo, Chris Bailey, Jim Ryder, Mata Haggis-Burridge, and Donatella Persico. 2019. "Games as (Not) Culture: A Critical Policy Analysis of the Economic Agenda of Horizon 2020." *Games and Culture*, June. https://doi.org/10.1177/1555412019853899.

Pritchard, Helen, Eric Snodgrass, and Magda Tyzlik-Carve. 2018. "Executing Practices." In *DATA Browser 06: EXECUTING PRACTICES*, edited by Helen Pritchard, Eric Snodgrass, and Magda Tyzlik-Carve, 9–24. London, UK: Open Humanities Press.

Raessens, Joost. 2006. "Playful Identities, or the Ludification of Culture." *Games and Culture* 1 (1): 52–57. https://doi.org/10.1177/1555412005281779.

Ruberg, Bonnie. 2015. "No Fun: The Queer Potential of Video Games That Annoy, Anger, Disappoint, Sadden, and Hurt." *QED: A Journal in GLBTQ Worldmaking* 2 (2): 108–24. https://doi.org/10.14321/qed.2.2.0108.

———. 2019. "Straight Paths Through Queer Walking Simulators: Wandering on Rails and Speedrunning in Gone Home." *Games and Culture*. https://doi.org/10 .1177/1555412019826746.

Ruberg, Bonnie, and Amanda Phillips. 2018. "Special Issue—Queerness and Video Games Not Gay as in Happy: Queer Resistance and Video Games." *Game Studies* 18 (3). gamestudies.org/1803/articles/phillips_ruberg.

Schüll, Natasha Dow. 2014. *Addiction by Design: Machine Gambling in Las Vegas*. Princeton, NJ: Princeton University Press. https://press.princeton.edu/titles/9156. html.

Søndergaard, Dorte Marie. 2002. "Theorizing Subjectivity: Contesting the Monopoly of Psychoanalysis." *Feminism & Psychology* 12 (4): 445–54. https://doi.org/10 .1177/0959353502012004006.

Staunæs, Dorthe. 2003. "Where Have All the Subjects Gone? Bringing Together the Concepts of Intersectionality and Subjectification." *NORA—Nordic Journal of Feminist and Gender Research* 11 (2): 101–10. https://doi.org/10.1080/080387 40310002950.

Walz, Steffen P., and Sebastian Deterding, eds. 2014. *The Gameful World. Approaches, Issues, Applications*. Cambridge, MA: MIT Press.

Walz, Steffen P, and Sebastian Deterding. 2015. "An Introduction to the Gameful World." In *The Gameful World: Approaches, Issues, Applications*, edited by Steffen P Walz and Sebastian Deterding, 1–13. Cambridge, MA: MIT Press. https ://doi.org/10.1007/978-3-319-00819-6.

Whitson, Jennifer R. 2014. "Foucault's Fitbit: Governance and Gamification." In *The Gameful World: Approaches, Issues, Applications*, edited by Steffen P. Walz and Sebastian Deterding, 339–58. Cambridge, MA: MIT Press.

Chapter 4

From Automating to Informating

Toward a Productive Model of Human/
Machine Collaboration in Higher Education

Jordan Canzonetta

As artificial intelligence is poised to change the technological, economic, and communicative landscape of the twenty-first century (Huws 2014; Markoff 2015; Reeves 2016), higher education will be inundated with educational technologies that aim to conduct teaching labor. Projections for advancements in educational technologies will likely outpace previous expectations for developing new tech and AIs.[1] Historically, in times of labor struggles and crises, employers deploy automation to compensate for insufficient human labor (McAllister and White 2006, 25). In higher education, this scenario manifests when educational technology is implemented to carry out time-consuming teaching tasks, such as standardized essay scoring or assessing student work (Hart-Davidson 2018, 248). This is true now more than ever as the pandemic has upended departmental labor structures and budgets. Labor in higher education is in crisis in the United States (Fulwiler and Marlow 2014), and thus the market is rife with technological solutions for pedagogical problems that are exacerbated by higher education's increased corporatization and reliance on contingent labor (Marsh 2004).

Many current models of educational technology are, to borrow Liza Potts's (2013) phrase, designed as "antisocial software," which are passed down to users who have little say in how the programs operate. This kind of software "automate[s] our work and [. . .] turn[s] [it] into a routine" without requiring any specialized or expert knowledge from humans (Grabill 2016). The difference between passively or proactively working with technology is what distinguishes unproductive and productive models of human/machine collaboration in higher education. It is also the difference between automating and informating human teaching labor. Automation is typically framed as a means of conducting teachers' skilled labor and doing their work for them instead of with them. Shoshana Zuboff's (1985, 1988) concept of "informating," however, offers a way to model automation to function "with" teachers; informating requires humans to make sense of the data and interactions between the human and machinic, which bolsters humans' expertise and ingenuity instead of deskilling them (Grabill 2015, 2016). This unique moment in history, when online education, automation, teachers, and economics are intersecting, offers real opportunities for scholars in HMC, composition and rhetoric, and online pedagogy to shape how automation and AI will influence communications-based education. This chapter aims to consider a productive model for how future collaborations between teachers and AI could be in the classroom. These considerations include what a "tight feedback loop between the machine and the human" could look like when humans and machines intentionally and productively work together in higher education contexts, especially on communicative labor, such as writing instruction (Markoff 2015, 212).

I compare two technologies, Eli Review (created by content specialists to informate) and Turnitin (created by nonexperts of writing to automate), to highlight how human/machine collaborations can flatten complexity in classrooms or encourage sophisticated human interactions and learning. The analyses from this chapter are based on five interviews from instructors who collaborate with Turnitin and five instructors who work with Eli Review (Beitin 2012). I assess these data via two content analysis methods for coding textual data: holistic and provisional coding (Saldaña 2009). My findings suggest informating can help teachers quickly identify pedagogical problems, encourage students' self-reliance, and carry out low-order managerial tasks to free up teachers' time for high-order tasks. Informating offers a productive model of human/machine collaboration that educators and designers could adopt as new and even more intelligent technologies emerge in higher education.

EXIGENCE

According to Jeff Grabill, co-creator of Eli Review, educational technology companies are actively trying to replace teachers. During his keynote speech at Zeeland Educational and Teacher's Academy, entitled "Robots are Coming," he indicated,

> For the last ten years of years of my life, I have spent plenty of time [. . .] at edtech investor meetings. Trust me, if you don't know this already, they are looking to replace you. You guys are messy. You get in the way. Robots are

clean[, they are] technologies that deprofessionalize teacher work versus technologies that professionalize your work, that require you [. . .] to invest your time and energy. (2015)

Such a claim is unsurprising, given that machines are often designed to either "replace" or "simulate" humans (e.g., artificial intelligence) or "augment" human ability (e.g., intelligence augmentation) (Markoff 2015, loc. 252). Automation acutely demonstrates a paradoxical dichotomy inherent in these approaches for technological design. While automation can extend human skill, it can also potentially displace us (loc. 233). After all, automation can replace human labor at a lower cost, and it frees employers from paying error-prone, unpredictable human workers (Reeves 2016, 153). When framed in this way, "the human is not valued for its ingenuity or creativity. Instead, when confronted with the cold efficiency of the machine, the human appears as just an organic collection of errors," which positions humans as a detriment to employers (153).

Automation is especially appealing to administrators, who have long been coping with financial constraints associated with the labor crises in higher education. Beginning in the mid-late 1970s, the academic labor force in the United States began shifting from secure, permanent tenure-track employment for instructional staff to a contingent model of hiring teaching labor (Schell and Stock 2001, 4). Contingent work in the larger U.S. economic landscape is typically categorized by "low pay and insecure work" and often "insufficient education and training" (Champlin and Knoedler 2017, 233); however, the latter is typically not applicable to contingent workers in academia, who are highly educated and work in an "industry with the highest proportion of contingent workers" in the country (232). Although percentages vary by institution type, the overall make-up of the academic workforce in the United States consists of over 70 percent non tenure-track academic positions, with a "62 percent increase in full-time non tenure-track faculty appointments and a 70 percent increase in part-time instructional faculty appointments" ("Annual Report" 2016) over the past four decades.

It comes as no surprise that educational technology programs' websites market time-saving, efficient labor to administrators whose staffs are stretched thin (Vie 2013a). Contingent laborers are working in such strenuous conditions that they crave algorithmic support—a welcome, though imperfect form of relief. For example, in a video hosted on the Turnitin website in 2015 entitled "Why Educators Love Turnitin," several instructors suggest the program enriches their quality of life by relieving the anxiety they feel about responding to an overwhelming workload. They use words, such as "lifesaver," "anxious," "upset," and suggest they could not "grade and evaluate" their students' work without Turnitin.

Rhetoric and composition scholars' attention to human and machine collaboration has centered around concerns about students' intellectual property, demonizing nonnative English speakers, inaccurate or one-dimensional assessment of student work, and creating tense dynamics between students and teachers in classrooms (Canzonetta and Kannan 2016; Canzonetta 2018; Howard 1999; Purdy 2005; NCTE 2013a, b; Vojak et al. 2011). Despite critique, these automated writing evaluation (AWE) programs and plagiarism detection services (PDSs) have flourished in the last two decades (Grabill 2016). For example, in 2012, Turnitin (the most globally popular PDS) was implemented in approximately 10,000 institutions and collected over 20 million papers; today, they serve over 15,000 institutions and have amassed over 929 million (and counting) student papers ("Homepage," Wayback Machine 2012; "Feedback Studio" 2018). AWE and PDS programs often market respite from pedagogical problems that arise from financial and labor issues in higher education (Marsh 2004; Vie 2013a, b). Criticism about popular educational technologies is often warranted and necessary, but it is not limiting their presence in higher education. A more proactive approach to resisting unethical or unproductive programs lies in working with those that balance and appropriately divide tasks between humans and machines.

AUTOMATING VERSUS INFORMATING: SITES OF ANALYSIS

One approach to forming these balanced collaborations is to adopt technology that needs teaching expertise to function appropriately. Technology, Zuboff claims, is frequently

> used only to automate production and thus reduce [human] skill and labor requirements. But its potential to inform organizational members about the work process and thus improve operations and increase innovation is the aspect of technology that will be most important to long-term organizational success. (1985, 5)

To informate in the context of educational technologies, then, requires the machine to function "with" teachers, so they can interpret data from student and machine interactions and translate it into pedagogy (Grabill 2016). For informating to happen, the design of the program must engage teachers in data interpretation that helps students improve their work before it reaches its "final" iteration. If automated data is used to assess student work or make a judgment call about its quality rather than helping students enrich their writing, it does not qualify as an informating technology. Informating provides teachers with the information they need to understand what actions to take in their lesson plans. Conversely, if automation is only deployed to grade students, it does not qualify as informed work because it reduces human skill (i.e., a teacher's expertise/training) needed to run a course. Put another way, automation is tied to assessing a student's final product, whereas informating is related to both the process of teaching and arriving at the product. Interpreting the process component of students' work requires teachers' expertise and encourages innovation, because instructors must develop creative and nuanced ways of working with data to teach students rather than simply assessing their work.

Automation has the potential to enhance human capacity and pedagogy, but many educational technology programs currently on the market are poorly framed, designed, and implemented. Automation becomes problematic when administrators and teachers expect it to comprehensively replace teachers' skilled labor. Current technologies cannot (yet) account for the specialized, adaptable, and intuitive processes teachers bring to their work. In most cases, automation is restrictive and bounded to parameters established by a designer. If students need help outside of those parameters, automation cannot appropriately respond. What automation can possibly offer, however, is facilitation of more human interaction between students, and it can address lower-order issues in classes that free up teachers' time to focus on their main pedagogical priorities.

Sites of Analysis

Turnitin and Eli Review are two technologies that represent the difference between automating and informating human labor in college/university writing classrooms.[2] Turnitin is an educational technology that was created over two decades ago; the program was designed to deter and find instances of student plagiarism through algorithmic detection. It automates teaching labor by finding matched text in students' work that could be plagiarized. Since its inception, the program has grown into a corporate "behemoth" and now offers writing assessment and artificial intelligence features that evaluate student work (Vie 2013b).

With a history spanning approximately two decades, Turnitin's features and technology have evolved heavily; the company has tested and expanded many tools aimed at bettering student writing. However, this chapter is focused on the plagiarism detection component along with one additional feature, Feedback Studio/GradeMark, an interface that connects teachers to students' "originality" scores and reports. Feedback Studio/GradeMark allows teachers to comment on student drafts (both textually and/or with audio feedback) and offers pre-labeled stamps for issues such as improper citation, run-ons, comma splices, lack of evidence, and so on ("What We Offer" 2016).

After students submit their work to Turnitin, teachers receive a report that offers an overall percentage score of matched text found in the document. Additionally, scores are available on a micro-sentence level with a link to the similarly identified text. Determining whether instances of matched text are plagiarized passages then requires a human reader, but that is not always clear to users. For example, a paper with an overall 0 percent score of matched text could be plagiarized and a paper with 50 percent matched text could be free of plagiarism. The percentages the algorithm produces are nondeterminate of plagiarism and necessitate instructors to attune to context and the ways in which students are writing, citing, and quoting—or not (Canzonetta 2018). However, because of Turnitin's widespread implementation through learning management systems, teachers are often exposed to Turnitin immediately and without opportunity to participate in training sessions, which could contribute to misinterpretation of the data Turnitin collates.

Eli Review was founded in 2011 by scholar-teachers Jeff Grabill, Bill Hart-Davidson, and Mike McLeod at Michigan State University's Writing, Rhetoric, & American Cultures department ("About Eli Review" 2018). The program was designed to be a "writing pedagogy—and learning theory—made into software," with the aim of bettering students' processes of revision and feedback (Hart-Davidson 2015). According to the originators of the program, Eli Review was formed as a reaction to peer-review processes that needed improvement in the creators' composition courses (Hart-Davidson 2015). Eli was designed by writing teachers to put pedagogical theories into practice in classrooms where students are writing ("Feedback and Revision" 2018).

The program is designed to focus instructors and students solely on three areas writing pedagogy: writing, review, and revision. The program does not offer features such as gradebooks/grading or plagiarism detection, and it does not allow teachers to comment directly on student documents. Teachers' feedback is limited to "endorsing" comments and providing students with a holistic textual overview at the very end of their review processes ("Eli Review Learning" 2020). Ultimately, I chose to study Eli Review and Turnitin, because they are familiar programs to writing instructors who have content knowledge in this area, and thus bend (and are bent by) the technologies in ways that nonexperts might not be.

METHODS

This chapter is a component of a larger project I conducted, where I adopted a mixed-methods new materialist comparative analysis of Turnitin and Eli

Review. I adopted two main methods for collecting data about human and machine engagements: Interviewing objects (Adams and Thompson 2016) and interviewing teachers who have worked with Turnitin and/or Eli Review (Beitin 2012) about their labor and experiences with these technologies in their classrooms. I used both sets of interviews to "follow" each technology and ask teachers questions about how they came into their professional lives, how they adapted them for their own purposes, and what role they play in their labor practices.

I assessed the interview data based on Saldaña's conceptualizations of holistic and provisional coding (2009). In the first round of coding, I searched transcripts for similar words and actions interviewees had responded with. After noting common phrases and topics among the respondents, I copied the transcripts and holistically color-coded them according to thematic similarities. In the second round, I coded data provisionally, which entailed using language from Adams and Thompson's object interviews methods to determine categories for coded data. I interviewed nine total instructors with various educational backgrounds (from R1s to community colleges) who were working at different levels of job security (ranging from graduate students, to full-time non tenure-track, to tenure-track professors), totaling five interviews for Eli Review and five for Turnitin because one participant had collaborated with both programs. Throughout the remainder of the chapter, I will refer to interviewees by their participant number to protect their anonymity (P1, P2, etc.).

Limitations

There were no part-time contingent faculty member interviews, perhaps because teachers who hold full-time positions at their institutions may feel more secure with time and job security to participate in research studies. Another constraint of this study is the limited degree to which two of the graduate students responded. Teachers who have been working longer have perhaps had more experience in reflecting about teaching practices and technological use. Instructors' positionalities also factored in as a limitation of the study. For instance, tenure-track participants were teaching no more than two classes a semester, which can, if they did not explicitly address their workload, affect the generalizations they were making about technological use and teaching. While these constraints do limit the scope of this study, the majority of participants' responses offer a rich qualitative look into how writing instructors from different backgrounds perceive their engagements with educational and learning technologies in their classrooms.

FINDINGS

Informating Makes Pedagogical Problems Visible Quickly

The first category of findings that emerged from the interviews relates to visibility—an idea instructors often used to describe how "informated" data about student work impacts their pedagogy and teaching labor. While Eli Review's team explicitly designed the program to informate teaching pedagogy, instructors used Turnitin data in similar ways and repurposed the tool's output to suit their pedagogical needs. Because they have content knowledge about best practices in writing, these instructors undermined Turnitin's original purpose for catching plagiarists and used data to teach students about attribution. It is important to note that this may not be the case for instructors who have little familiarity with writing pedagogy, and Turnitin caters to instructors across all disciplines. Rather than using autogenerated data to assess student work, Turnitin teachers used data to quickly find areas in which students struggled with citation practices. In doing so, both Eli Review and Turnitin teachers use the tools to teach students more about writing rather than to castigate them for errors. Data, thus, is providing teachers a way to informate teaching labor because it provides quick and targeted information about where students struggle individually and as a group. With this information, teachers then design targeted lesson plans that are based on students' specific problems in class.

For instance, Eli Review's simple design approach for collating student comments and reviews helps one instructor, P1, understand the common threads or issues students are dealing with in class. As she remarks, "looking at the comments [students] write has given me clues to the things that they're struggling with," which is particularly true when those comments are easily comparable, because they appear next to each other. P2, who uses Turnitin, indicates a like perspective and asserts the program helps him "flag major things" or areas where students struggle, especially concerning patchwriting, paraphrasing, and summarizing. Furthermore, he believes the software has "made him pay more attention to how [his] students use sources," where their information comes from and how they use research in their writing.

P2 is careful not to assess student work based on the findings the machine produces. P7 also has to resist requests from teachers or administrators who want a clear and definitive "threshold" similarity index number that indicates if a student has plagiarized. P2 notes that Turnitin's "similarity index" cannot accurately find plagiarism and can in some cases miss it. In one case, he received a student paper with "88% [matched text] that wasn't plagiarized [and] a case of plagiarism with a similarity index of 11%." Instead, he uses similarity scores to signal where he needs to "check" student work to see if they need additional help. Similarly, for P6, Eli Review helps her target not only where students are "struggling" but also how she can help them through modeling responses with peers.

On a larger, organizational level, both tools enhance visibility through their interfaces and the ways in which automated data is presented. For P3, Turnitin quickly helps her see where students are getting information for their research papers; it is "easy" for her to locate where their cited information comes from because of the color coding and convenient pathway it offers to the original source. For P9, Eli Review's features and interface allow him to "document" student writing over the course of a semester, which illuminates "all of the work that goes into a [student's] writing process," thus saving him time on the low-order managerial task of collecting those materials himself.

Teaching alongside these technologies means instructors are working to develop new pedagogical approaches, while also functioning as a buffer and translator between the machine and students. Brock and Shepherd (2016), Reyman (2017), and Ingram (2013) argue algorithms are rhetorical because the implicit and hidden parameters through which they function persuade users toward particular engagements and actions. But in this (arguably, best scenario) case, the visibility of automated information persuades instructors to change their teaching practices to meet student needs. The instructors from the human interviews have writing pedagogy background and experience, which frames how they implement those changes and interpret automated and informated data. In the case of Turnitin, automated data that is meant to assess is instead transformed into informated data because of how teachers use the data to teach. Instructors mediate between the programs' designs, required pedagogical outcomes, and desired student learning to align with their goals for students and restructure their roles and relationships with the machines, shifting their agency and negotiating it (Cooper 2011). These instructors with experience teaching writing have a fluid approach to working with technology, because they understand the impacts certain design features can have on students' learning.

Convenience and Low-Order Teaching Priorities Are Key

In terms of teaching labor, convenience is by far the most common way in which instructors describe how Turnitin or Eli Review impacts their labor. Convenience is framed in terms of time saved on organizational labor, time saved on finding pedagogical problems students are having, and, in general, as a means of simplifying low-order teaching tasks that are associated with the managerial component of running a writing course. The managerial layer of teaching writing, Strickland (2011) argues, is often underrepresented in scholarship and has negative connotations; thus, it follows that managerial labor is low priority for writing instructors.

For teachers who adopt Turnitin in their classrooms, the platform is a way for instructors to "easily" distinguish when students have purchased papers, passed their work along to other students, or copied and pasted entire articles as their own work (P1, P2, P3, P7, P8). Rather than manually typing out passages, students may have plagiarized (or copying and pasting text) into a search engine to scour the Internet for copied words, Turnitin's algorithms cut out the middleman and do the labor of searching for teachers. P1 describes this process as a convenience that saves her time; prior to Turnitin, she "used to spend an hour on a suspicious essay" which she indicated was a task that she considered low priority. Automation can, according to Kennedy, "handle [menial] tasks more efficiently, more consistently," which allows humans to work on "higher-order concerns" (2016, 118). The program helps P1 "confirm whether plagiarism has taken place pretty quickly," without requiring her to investigate student work on her own. Determining plagiarism is not one of P1's main concerns in her classroom; it is a low-order task that Turnitin's algorithms can help her manage so that she can spend time on commenting on students' revision plans, which is a critical part of her pedagogy. All of the other writing instructors who worked with Turnitin expressed a similar sentiment, suggesting that Turnitin is an ancillary tool that helps teachers locate the "knucklehead" attempts at plagiarism quickly (P2, P3, P7, P8).

Another instructor also pointed to an unexpected convenience Turnitin offers, which supports his administrative labor when he must document academic integrity breaches at his institution. In any case of student-based academic integrity infractions, instructors must submit proof to the institution of plagiarism, for which he prints the Turnitin report. This type of administrative management is a low priority for P2. It is important to note that framing Turnitin reports as hard "proof" of plagiarism, even though they are fallible, legitimizes the tool to administrators. P2 is vocal about understanding the limitations of Turnitin and notes that it cannot actually catch plagiarism because a human must do that contextual interpretation.

Even if teachers are successfully repurposing the algorithmic functions to serve a pedagogical aim, the Turnitin company still stores students' intellectual property and surveils them (CCCC-IP Caucus 2006; NCTE 2013a; Zwagerman 2008). Collaborations between teachers and Turnitin are complicated because they are layered and have implications within institutional and corporate settings. Rather than working with Turnitin to monitor and punish all potential student plagiarists, which is what the program was initially designed to do, the teachers I interviewed adapt it to find severe instances of plagiarism in their courses and talk to students about them. All instructors who engaged Turnitin noted that their main goals are not to punish students, but to teach them about proper attribution. In some cases, institutional mandates, as is evident with the example P2's documentation process, may take precedent over teachers' pedagogy and course aims.

Instructors who collaborated with Eli Review share similar responses about convenience: Eli helps them conduct low-order administrative or managerial labor that saves them time. For P9, Eli Review allows him to "archive" previous assignments that he has fine-tuned and developed for his students. The application allows him to "reuse" them and carries out menial tasks by "doing some really efficient things like assigning the same activity to two different classes or assigning the same activity to 100 different classes." This shortcut allows P9 to save his energy and decreases his "intellectual labor" for old assignments, enabling him to spend more time focusing on individual students. P5 expresses a similar perspective, where she claims Eli Review helps her search and log student work to "keep track of all the pieces" of the writing process in her courses.

P1 suggests Eli Review helps her facilitate organizational labor, because the labor she conducts in her classroom is modeled from the pedagogy that is built into its design. P9 and P5 view Eli Review as centralizing and streamlining elements of their organizational labor that can be done with other tools. Blackboard (a course management system) offers similar functionality, but it is more difficult to use and is less efficacious than Eli Review (or as P5 put it, using Blackboard, to do so is "a bloody mess"). However, P1's perception of Eli Review is such that she feels as if she would struggle with her time and labor if she did not have access to the program:

> If I were to replicate the things I do on Eli on paper, the workload would be enormous. It would be huge; it would be so huge that I wouldn't do it. Which was probably why. . . . I wasn't doing those things before. . . . Eli just makes some of those good teaching activities just so much easier and faster, and instantaneous. . . . Instead of reading through 30 sheets of paper I can just pull up the screen and I can see things within 30 seconds . . . it's made some of the brain work—the brain labor a little bit easier.

P1 suggests Eli Review helps her save time and energy, which makes high-priority teaching tasks and intellectual labor "easier" for her. Furthermore, Eli Review altered P1's pedagogy; she models her lesson plans and classroom activities on the program; thus, it logically follows that her workload would intensify without the tool. Now that Eli Review has prompted changes in her pedagogy, she is more tied to the program and reliant on it. Without the program, her labor load would increase, and she might have to sacrifice other high-priority teaching tasks to accommodate her workload.

These low-order administrative and managerial tasks Turnitin and Eli Review carry out are appropriate and balanced uses of automation in writing classrooms, because they are not conducting assessments. Rather than automating human work and deskilling instructors, informating offers a means of centering teaching expertise and course management skills; this is a clear example of Zuboff's (1985, 1988) vision of informating, which enriches pedagogy and student learning. Machines cannot account for context or adequately assign grades to student writing, because they cannot respond to information outside of their algorithmic parameters.

Furthermore, language is complicated and fickle; machines cannot (yet) capture meaning, tone, or arguments that students are making in their writing; they only work to standardize and reinforce low-order priorities such as grammar and spelling (Vojak et al. 2011). Although Turnitin is not specifically designed for the ways in which teachers are adapting the program, they are modifying the tool's algorithmic output to work best for them and their pedagogical goals. Eli Review is more intentionally designed to carry out low-level managerial tasks for teachers in their classrooms and has the added benefit of built-in pedagogy designed by content experts. Such a model functions to extend and augment human labor rather than displacing it, as Markoff would suggest (2015). However, it is important to note that instructors felt Eli Review increases instructors' labor as they begin to learn the program because it requires them to reframe their teaching approach, to look at student writing more frequently, and teachers have to learn how to use the quantitative data the program generates (P1, P4)—something that could be a particular burden to precariously employed contract academic workers.

Self-sufficiency and Human-to-Human Interaction Increases

For both Turnitin and Eli Review, the idea of working with technology to help students strengthen their independence echoes throughout the interview data, which is illustrated in several passages below. Instructors' objectives for more "self-sufficiency" among students were twofold: self-sufficiency functions both as a means of reducing teaching labor and as a pedagogical imperative to encourage students to collaborate (P1). Overwhelmingly, instructors wanted students to cultivate more confidence in their writing, and they wanted to help students learn from each other to mitigate the labor of low-order teaching tasks (P1, P2, P4, P5, P6).

For instance, P2 asserts that part of his teaching labor as a first-year writing teacher is to "de-freshmanize" students by acting as both instructor and "counselor" to students. He feels that part of his labor is devoted to teaching students how to participate in the university, which constrains the time he devotes to higher-order pedagogical aims. If students were more self-sufficient, P2 could spend more time on crafting content for class. He expresses frustration with the unintended labor he feels responsible for in his classroom. Such labor detracts from the course content P2 wants to teach and puts his classes behind schedule. For him, Turnitin offers a means of "knocking off the naiveté" first-year students bring to his classroom, because it helps them learn how to be responsible for their own academic integrity and frees up some of his time. He allows students to see how Turnitin presents data about their work, he teaches them how to understand pieces of it, and then offers them a chance to revise their citation practices so he can spend time on other subject matter in class.[3]

Another instructor mentions students' self-sufficiency as it relates to (1) their "independence" as writers and (2) her own labor of commenting on student work quickly enough for students to revise it while their work is fresh in their minds. P5 sought out Eli Review because she was worried that students were too reliant on her and were dismissive of peer comments. She mentions, "You have one instructor, 22 students in a class; it takes a long time for them to get that feedback and . . . they're not responding to that initial timely feedback that we want them getting." In working with Eli Review as a lever to shift some of the labor of reviewing student work to students, P5 hopes students will become more confident in their writing, both as they learn how to provide strong feedback and as they frequently revise their work. Doing so frees up time for P5 to focus on higher-order teaching tasks by eliminating the repetitiveness she writes in her feedback to students. For instance, rather than repeating how to address a common mistake or problem many students have in their writing (P5 uses the example of student issues with MLA formatting), P5 wants students to "self-evaluate" and use Eli Review to create accountability among peers. She notes that "the structure of the peer review and the revision plan puts that burden on the students" because students have access to seeing each other's comments and the instructor's endorsements, which helps them model strong peer review and keeps P5 from having to repeat her comments frequently.

The paradox Markoff refers to regarding machines replacing humans or augmenting them could be applied to Eli Review in this capacity (2015, loc. 232). Does augmenting students' capacity for self-sufficiency replace the work teachers do in writing classrooms? I argue no; students still need instructors' input in Eli Review to model and endorse the kind of peer-reviewing they should strive to achieve. Eli Review enhances their ability to become self-sufficient because a teacher's formative feedback for one student can be seen by all students, thus strengthening their ability to determine strong practices for revision. Doing so frees up instructors' time on commenting on students' peer revisions so they can work on other teaching priorities. This results in students' bolstered sense of confidence and collaborative abilities with each other, which positions the teacher as a coach or guide who is still a necessary part of the classroom.

As the instructors from this study indicate, students' self-sufficiency is highly tied to how instructors implement and frame each program to their students. Transparency about the programs' capabilities and limitations is an important factor in encouraging students to have beneficial partnerships with machines, as is offering students a chance to learn from data and revise their work. Modeling student work and endorsing it in a public way teaches students how to learn from each other and encourages collaboration among peers. Students can develop stronger senses of confidence and independence when the machines show them where they need help or when other students are doing good work, which targets and focuses students' attempts at revision. Good practice, then, requires teachers to understand informed data as a crucial part of students' revision processes. Thus, technology has the potential to augment the class experience for both teachers and students in helping the latter become more self-reliant; students gain confidence in their writing and teachers are able to spend their time on high-priority teaching tasks.

EMBRACING INFORMATING

Collaborations with Eli Review and Turnitin have several implications tied to labor and pedagogy for both teachers and students. As regards pedagogy, when instructors work closely with these technologies, they appear to act as buffers between student and machine, reworking, undermining, or collaborating with algorithms to ensure students are benefiting from working with the programs. Turnitin is not designed to informate data about student work; however, some teachers allow students to submit multiple drafts of their work to learn from text-matching algorithms to revise their writing or instructors design lesson plans based on the problems they see arising from automated data. For Eli Review to work, students must work together and learn from each other as they construct peer reviews in class while an instructor uses her expertise to publicly model and endorse strong writing practices to the whole class. Even though there are design constraints for both programs, teachers from these interviews (who, again, are experts of student writing) manipulate the algorithmic functions to best suit their students' needs or their own classroom objectives.

Visibility also overlaps with self-sufficiency in the data from interviews with instructors. Eli Review encourages students to collaborate with their peers, and the program's endorsements and modeling functions make visible strong instances of writing, which encourage students to follow each other's examples rather than following a machine's example. Turnitin also makes students' citation needs more visible because the program signals where students may be struggling; if given the opportunity, they can use those signals to revise their writing. There are, of course, implications to having such intimate insight into students' work. Collaborating with machines to gather insights into students' writing processes also means the designers of the programs have access to the same data. While Eli Review requires IRB approval for the studies they conduct based on student work, Turnitin is not beholden to the same standards because it does not have ties to a particular educational institution. Thus, there are no ethical checks or balances for the work Turnitin generates about student users, which is all the more complicated by their intellectual property practices, corporate ties, and sometimes harmful consequences for student writers (CCCC-IP Caucus 2006; Canzonetta and Kannan 2016; Canzonetta 2018; Howard 1999; Marsh 2004; Vie 2013a, b, 2017). Additionally, on a micro level, surveillance of student work can also encourage teachers to micro-manage their work and get too bogged down in the quantitative data about their writing processes, which could lead them to spend more time "in the weeds" of student writing rather than focusing on high-order teaching concerns related to content (P1).

Labor and pedagogy are often tied to each other in this analysis; however, there were a few distinctions between the two categories that emerged in the analysis. Labor and machines were often discussed in terms of convenience and the managerial labor associated with running a writing classroom. Instructors collaborated with Turnitin and Eli Review not to assess student work, but to use information as a means of signaling where students struggled, which saved them time in some respects. Instead of trying to discover where students needed help by sorting through papers and collating themes or categories that emerged, the informed data eliminated an extra step for instructors and conducted the labor of locating information. Working closely with machines in this context helps instructors because it allows them more time to tailor specific lesson plans that address the issues students are having difficulty understanding and keeps humans in the "tight feedback loop with machines" mentioned earlier (Markoff 2015).

Eli Review requires instructors to frontload their labor at the beginning of their courses as they experience the learning curve for working with the program. Regardless of their teaching loads, instructors in our sample accepted the additional labor that sometimes accompanies working with Eli Review because they see such a significant pay off for their students. Eli Review augments student writers and teachers' instruction by extending human capacity rather than automating teaching labor to expedite the writing process. Turnitin is described as a moderate convenience for teachers who engage with the program, but there are tradeoffs for some teachers, who worry about how other instructors may collaborate with Turnitin to punish students, among the ethical concerns listed above.

In future, collaborations with machines in higher education must work toward a productive balance between teachers and educational technologies. Often, teachers do not have a say in what technology is passed down to them from administrators, so they find ways to undermine the tools and make them work "with" their students (which was the case for several instructors from this study who worked with Turnitin). Even so, the algorithmic functions of the programs still "persuad[e] users into particular engagements" and coerce teachers into changing their modes of instruction—sometimes for the better, sometimes not (Brock and Shepherd 2016, 17). Automating data about student labor does not have to be one-dimensional or detrimental in classrooms nor does it have to deskill teachers. If we instead frame the way automation can informate work and keep human expertise at the center of educational labor in a post-pandemic world, productive and balanced collaborations can happen. As more intelligent technologies infiltrate higher education, informating can both help safeguard teachers' job security, and it can enhance and extend their pedagogy.

NOTES

1. This is especially so with educational technology companies profiting significantly from the COVID-19 pandemic, allowing them to allocate more resources for developing new tech and different forms of automation which will target pandemic-related classroom issues (Dolan 2020).

2. Examples of Turnitin's interfaces are located on their website at www.turnitin .com; examples of the interfaces for Eli Review are available at www.elireview.com. Screenshots of these images were not included due to permissions concerns.

3. This algorithmic process could be seen as the gamification of their own writing practices, with the awareness of quantified matching text scores leading students to try to get "better" scores. This form of algorithmically led behavior is also discussed, with a critical eye, by Gómez, chapter 3, this book. — Ed.

REFERENCES

"About Eli Review." November 21, 2018, http://elireview.com/about/.

Adams, Catherine and Terrie Lynn Thompson. 2016. *Researching a Posthuman World: Interviews with Digital Objects*. London, UK: Macmillan Publishers Ltd.

"Annual Report on the Economic Status of the Profession, 2015–16." 2016. *Academe* 102(2), 1.

Beitin, Ben K. 2012. "Interview and Sampling: How Many and Whom." In *The Sage Handbook of Interview Research: The Complexity of the Craft*, edited by Jaber Gubrium et al. 243–252. Thousand Oaks, CA: Sage Publications.

Brock, Kevin and Dawn Shepherd. 2016. "Understanding How Algorithms Work Persuasively through the Procedural Enthymeme." *Computers and Composition* 42, 17–27.

Canzonetta, Jordan and Vani Kannan. 2016. "Globalizing Plagiarism & Writing Assessment: A Case Study of Turnitin." *Journal of Writing Assessment* 9(2). http://journalofwritingassessment.org/article.php?article=104.

CCCC-IP Caucus. 2006. "Recommendations Regarding Academic Integrity and the Use of Plagiarism Detection Services." https://culturecat.net/files/CCCC-IPposi tionstatementDraft.pdf.

Champlin, Dell P and Janet Knoedler. 2017. "Contingent Labor and Higher Education." *Review of Political Economy* 29(2), 232–248.

Cooper, Marilyn M. 2011. "Rhetorical Agency as Emergent and Enacted." *College Composition and Communication* 62(3), 420–449.

Dolon, Mike. 2020. "Column: Big Funds Circle EdTech as Post-Pandemica Mega Trend." *Reuters.* https://www.reuters.com/article/us-global-education/column-big -funds-circle-edtech-as-post-pandemic-mega-trend-idUKKCN26G16P.

"Eli Review Learning Resources." February 27, 2020, https://elireview.com/learn /tutorials/debriefing/endorsement/.

"Feedback and Revision." April 21, 2018, https://elireview.com/content/td/ feedback/.

"Feedback Studio." March 31, 2017, https://turnitin.com/en_us/what-we-offer/feedback -studio.

Fulwiler, Megan and Jennifer Marlow. 2014. *Con Job: Stories of Adjunct and Contingent Labor.* Computers and Composition Digital Press/Utah State University Press. https://ccdigitalpress.org/conjob ,

Grabill, Jeff. 2015. "Robots are Coming: Technologies are Changing the Teaching of Writing." *Zeeland Educational and Teacher's Academy,* https://www.youtube.com /watch?v=6B8LtNRN6kw.

———. 2016. "Do We Learn Best Together or Alone? Your Life with Robots." *Computers & Writing Conference,* http://elireview.com/2016/05/24 /grabill-cw-keynote/.

Hart-Davidson, William. 2015. "The Power of Peer Learning." Presented at the Conference on College Composition and Communication. Tampa, FL, https:// elireview.com/2015/03/13/cccc-ignite/.

———. 2018. "Writing with Robots and Other Curiosities of the Age of Machine Rhetorics." In *Routledge Companion to Digital Writing & Rhetoric,* edited by Jonathan Alexander and Jacqueline Rhodes. 248–256. Abingdon, UK: Routledge.

"Homepage." June 13, 2016, http://turnitin.com.

Howard, Rebecca Moore. 1999. *Standing in the Shadow of Giants: Plagiarists, Authors, Collaborators.* Stamford, CT: Ablex.

Huws, Ursula. 2014. *Labor in the Global Digital Economy: The Cybertariat Comes of Age.* New York: Monthly Review Press.

Ingraham, Chris. 2013. "Toward an Algorithmic Rhetoric." In *Digital Rhetoric and Global Literacies: Communication Modes and Practices in a Networked World,* edited by Gustav Verhulsdonck and Marohang Limbu. 62–79. Hershey, PA: IGI Global.

Kennedy, Krista. 2016. *Textual Curation: Authorship, Agency and Technology in Wikipedia and the Chambers' Cyclopedia*. Columbia, SC: University of South Carolina Press.

Markoff, John. 2015. *Machines of Loving Grace: The Quest for Common Ground between Humans and Robots* (1st ed.). New York, NY: Ecco, an imprint of HarperCollins Publishers. Kindle.

Marsh, Bill. 2004. "Turnitin.com and the Scriptural Enterprise of Plagiarism Detection." *Computers and Composition* 21, 427–438.

McAllister, Ken S. and Edward M. White. 2006. "Interested Complicities: The Dialectic of Computer-assisted Writing Assessment." In *Machine Scoring of Student Essays: Truth and Consequences*, edited by Patricia Freitag Ericsson and Richard Haswell. 8–28. Logan, UT: Utah State UP.

NCTE. 2004. "CCCC Position Statement on Teaching, Learning, and Assessing Writing in Digital Environments." https://cccc.ncte.org/cccc/resources/positions/digitalenvironments.

———. 2013a. "Resolutions & Sense of the House Motions." *NCTE.org*, http://www.ncte.org/cccc/resolutions/2013.

———. 2013b. "Machine Scoring Fails the Test: NCTE Position Statement on Machine Scoring." *NCTE.org*, http://www.ncte.org/positions/statements/machine_scoring.

Potts, Liza. 2013. *Social Media in Disaster Response: How Experience Architects Can Build for Participation*. Abingdon, UK: Routledge.

Purdy, James. 2005. "Calling Off the Hounds: Technology and the Visibility of Plagiarism." *Pedagogy* 5(2), 275–296.

Reeves, Joshua. 2016. "Automatic for the People: The Automation of Communicative Labor." *Communication and Critical/Cultural Studies* 13(2), 150–165.

Reyman, Jessica. 2017. "The Rhetorical Agency of Algorithms." In *Theorizing Digital Rhetorics*, edited by Aaron Hess and Amber Davisson. 112–125. Abingdon, UK: Routledge.

Roberts, Katherine. 2014. *Convenience Sampling through Facebook*. Thousand Oak, CA: Publications.

Saldaña, Johnny. 2009. *The Coding Manual for Qualitative Researchers*. London: Sage Publications.

Schell, Eileen E. and Patricia Lambert Stock, P. 2001. *Moving a Mountain: Transforming the Role of Contingent Faculty in Composition Studies and Higher Education*. Urbana, IL: National Council of Teachers of English.

Strickland, Donna. 2011. *The Managerial Unconscious in the History of Composition Studies*. Carbondale: Southern Illinois University Press.

Vie, Stephanie. 2013a. "A Pedagogy of Resistance toward Plagiarism Detection Technologies." *Computers and Composition* 30, 3–15.

———. 2013b. "Turn it Down, Don't Turnitin: Resisting Plagiarism Detection Services by Talking about Plagiarism Rhetorically." *CCC Online Journal*, http://cconlinejournal.org/spring2013_special_issue/Vie/ .

———. 2017. "Plagiarism Detection Services Are Money Well Spent." In *Bad Ideas about Writing,* edited by Cheryl Ball and Drew M. Loewe. 287–293. Morgantown, WV: Digital Publishing Institute.

Vojak, Colleen, Sonia Kline, Bill Cope, Sarah McCarthey, and Mary Kalantzis. 2011. "New Spaces and Old Places: An Analysis of Writing Assessment Software." *Computers and Composition* 28(2), 97–111.

"What We Offer." June 10, 2016, http://turnitin.com/en_us/what-we-offer/.

"Why Educators Love Turnitin." 2012. https://vimeo.com/40676105.

Zuboff, Shoshana. 1985. "Automate/Informate: The Two Faces of Intelligent Technology." *Organizational Dynamics* 14(2), 5–18.

———. 1988. *In the Age of the Smart Machine: The Future of Work and Power*. New York: Basic Books.

Zwagerman, Sean. 2008. "The Scarlet P: Plagiarism, Panopticism, and the Rhetoric of Academic Integrity." *College Composition and Communication* 59(4), 676–710.

Part 2

ROBOTS AND SOCIAL JUSTICE

Chapter 5

The Misogyny of Transhumanism

Nikila Lakshmanan

HUMANISM, TRANSHUMANISM, AND POSTHUMANISM

Humanism, transhumanism, and posthumanism are social movements that have captured the imagination of Silicon Valley for decades and continue to fascinate scholars in the field of feminist theory. In "Posthumanism, Transhumanism, Antihumanism, Metahumanism, and New Materialisms: Differences and Relations," the feminist philosopher Francesca Ferrando (2013) explains that humanism is a philosophical tradition that originated in the Enlightenment era (26). Optimistic about the human capacity for eternal progress, humanists perceive the human individual as a rational, autonomous, and exceptional being superior to all other species in the global ecosystem. Humanism has long been the subject of feminist criticism, because it has historically seen the default human as an abled, wealthy, and cisgender white man (26–27). Humanism dominated Western thought in various forms from the Enlightenment to the end of the twentieth century.

Transhumanism and posthumanism both emerged in response to the waning of humanism, gaining momentum in the twenty-first century. Both movements seek to replace humanism. As a result, posthumanism and transhumanism are often mistaken for one another. The confusion stems not only from their similar names but also the fact that their adherents share a perception of "the human" as a malleable construct. Both seek evolution beyond the human and use the word "posthuman" in reference to their ideals (Ferrando 2013, 26–28). However, it is important to note that transhumanists and posthumanists are distinct philosophical traditions with very different conceptions of the posthuman.

The focus of this chapter is transhumanism, but first it is worth defining posthumanism.[1] Ferrando has argued that posthumanism stemmed from post-modern feminist theory in the 1990s. She notes that posthumanism originated in the field of literary criticism and spread to critical theory, cultural studies, and philosophy. Posthumanists inherited the postmodern feminist rejection of humanism. They claim that humanism has never really seen women as fully human. They also maintain that humanism is behind the dualistic think-ing in Western thought they see as the source of sexism (Ferrando 2013, 26–29). Posthumanists hold that dualistic thinking draws essentialist binaries between reason and emotion, mind and body, and manhood and woman-hood, always privileging the former over the latter. Ferrando's account of posthumanism is echoed by the philosophers Irina Deretić and Stefan Lorenz Sorgner in the introduction of their book *From Humanism to Meta-, Post-, and Transhumanism?* (Deretić and Sorgner 2016, 14–18). They recall that the term posthumanism was coined in 1977 by the literary critic Ihab Hassan and explain that the movement is closely linked to the tradition of continental philosophy, placing a strong emphasis on non-dualist ways of thinking and acting (13).

As Sorgner, Deretić, and Ferrando all suggest, humanism promotes the notion that men are inclined to reason, while women are inherently emotional and less capable of the rational thought to which real humans are supposed to have a predisposition (Ferrando 2013, 28; Deretić and Sorgner 2016, 14–15). Humanism privileges men over women and cultivates the ideology of misog-yny, which the philosopher Kate Manne defines in her book *Down Girl* as the hostility to and contempt for women needed to sustain systems of sexism and patriarchy (Manne 2017, 43). In response to misogyny, the posthumanist conception of the posthuman is that of citizenship within a social utopia: a world that has left behind the oppression of women and makes room for all life forms historically excluded from the humanist definition of humanity. Ferrando, Deretić, and Sorgner suggest that posthumanists welcome tech-nologies that can be used to spread feminist ideas and resist the oppression of women (Ferrando 2013, 28; Deretić and Sorgner 2016, 14–15).

However, Deretić and Sorgner also point out that posthumanism is less unified and more ambivalent on science and technology than transhumanism (2016, 14–15). In his article *What is Transhumanism?*, the transhumanist James Hughes has suggested that many strains of posthumanism—unlike transhumanism—belong to a larger family of critical theory and practice that has sometimes been called left-wing bio-conservatism. Compared to transhu-manists, posthumanists are more cautious in their approach to enhancement technology (Hughes 2005, 9). Posthumanism is concerned with techno-social rather than purely technological development. Posthumanists and other left-wing bio-conservatives worry about the implications that enhancement technologies have for women and are skeptical of the notion that technology can help human beings transcend the limitations of their bodies.

THE FOUR CAMPS OF BIOPOLITICS

A careful assessment of the literature on transhumanism shows that of all biopolitical camps to weigh in on the subject, four have been particularly influential. These camps are libertarian and democratic transhumanism, as well as left and right-wing bio-conservatism. Right-wing bio-conservatives stem from Christian and Jewish theological roots of bioethics, and they argue that human beings should cede procreative and evolutionary matters to God, gods, or a causal force in a god-like role (e.g., "Nature"). In her book *The Bioethics of Enhancement*, the feminist philosopher Melinda Hall observes that right-wing bio-conservatives, such as the Jewish philosopher Leon Kass, call on human beings to protect the boundaries of human nature and dignity from technological interference and temptation (Hall 2017, 6). These bio-conservatives are popular among Christians in the Religious Right in the United States, even if they are not Christian, and they endorse the notion that traditional gender roles are intrinsic to humanity—including Kass himself in his article "Man and Woman: An Old Story" (Kass 1991, para. 3). Left and right-wing bio-conservatism rarely interact with one another. Ironically though, some overlap exists between right-wing bio-conservatism and trans-humanism in the way they draw on the legacy of humanism: both see the human as the central player in the global ecosystem.

Hughes has argued that transhumanism has split into two distinct strains of thought: democratic and libertarian transhumanism (Hughes 2002, para. 3–5). Transhumanism is a technophilic, futurist movement that calls for using enhancement technology to upgrade the human into beings with physical and cognitive capacities so far superior to those of the humanist conception of humanity that one could no longer consider them the same species. This technological utopia is what *transhumanists* call "the posthuman" (para. 3–5). Transhumanists envision the posthuman as everything from uploaded minds and grey matter "copied" onto silicon, to synthetic artificial intelligence, and comicbook-like superhumans who are stronger, faster, smarter, and more conventionally attractive than humans could ever hope to be (Hall 2017, 11, 85). However, there is one big difference between democratic and libertarian transhumanism. Democratic transhumanists call for an equal access to enhancement technology, which could otherwise be limited to privileged sociopolitical classes and related to economic power, thereby reinforcing systems of oppression (Ferrando 2013, 27). Libertarian transhumanism, on the other hand, is popular in the alt-right and other far-right political movements in the United States; it advocates an unbridled free market as the best guarantee of the right to human enhancement.

The friction between posthumanism and transhumanism stems primarily from the latter's libertarian strain, which dominates the mainstream transhumanist movement. Misogyny has poisoned transhumanist approaches to enhancement technologies, especially reproductive biotechnologies, which seem to have imbibed the humanist tradition of seeing women as less human (Hall 2017, 69–71). The far-right in the United States has embraced transhumanism because much of its logic is linked to eugenics. This chapter will demonstrate the misogyny of mainstream transhumanist approaches to the following technologies: in-vitro fertilization (IVF), preimplantation genetic diagnosis (PGD), artificial wombs, and sex robots. As Hall explains, IVF is a medical procedure through which a sperm fertilizes an egg via artificial insemination, and the zygote is implanted artificially in the womb for gestation and birth. PGD is a biotechnology with which potential parents screen the embryos they create through IVF for various traits, from congenital diseases to sex, before choosing the ones they will implant in the womb (Hall 2017, 86). The transhumanist Zoltan Istvan notes that artificial wombs are synthetic wombs that can gestate embryos formed through IVF (Istvan 2019, para. 3). Istvan also notes that sex robots represent a range of AI technologies designed to enhance sexual pleasure (Istvan 2014b). I cite Istvan and other libertarian transhumanists within and outside academia to show that their unconditional endorsement of these technologies ignores the risks they pose for women.

However, unlike Hall, I do not believe the tradition is irredeemable (Hall 2017, 86). In the final section of the chapter, I make recommendations for how transhumanism can move past its problem of misogyny. Transhumanism must abandon libertarianism and embrace its democratic strain, which represents the movement's best instincts for social justice regarding the use of enhancement technologies. Democratic transhumanism has been overshadowed by libertarian transhumanism; with it, posthumanism in the feminist sense has, like other strands of left-wing bio-conservatism, been eclipsed by its right-wing varietal. Democratic transhumanism has the potential to engage in a fruitful dialogue with a broader posthumanism. Only when the two movements meet can they offer a viable alternative to the regressive gender politics of both right-wing bio-conservatism and libertarian transhumanism.

TRANSHUMANISM AND FAR-RIGHT POLITICS

Transhumanism, as I mentioned earlier, seeks to push the boundaries of humanism by challenging the concept of human exceptionalism.

Transhumanists exalt in the utopian possibility that humans can one day merge with machines. Ironically, Ferrando (2013, 27) observes that transhumanism is also sometimes recognized as a kind of "ultra-humanism." Although they are dissatisfied with the limitations of the human condition, transhumanists see the human as uniquely positioned for their lofty metaphysical goals, and they apply the humanist zeal for eternal progress to the goal of transcending the human itself. Feminist theorists like Hall have denounced transhumanism on the grounds that it continues the humanist tradition of failing to recognize the humanity of women (Ferrando 2013, 27–28; Hall 2017, 6, 69–71).

It is undeniable that transhumanism has regressive roots. In 1998, the World Transhumanist Association (now called Humanity+) penned "The Transhumanist Declaration" (Baily et al. 2009) as its founding document. The Declaration calls on humanity to invest time and research into technologies to improve cognition, antiaging techniques, cryogenics, and reproductive biotechnology and argues that human beings will gain more control over their lives if they have a better understanding of how to use technology to improve their physical and cognitive capacities (Baily et al. 2009, para. 1–8). On the surface, this vision may not seem to contradict feminism. However, Hall is disturbed by the "Declaration"'s keenness to eradicate all traces of human vulnerability and dependence, especially with respect to aging, disease, and "feeble" memories and intellects (Hall 15, 109). She warns that the transhumanist ideology stigmatizes bodies that do not conform to the rational, autonomous ideal it inherited from humanism (6, 11). She suggests that transhumanism fails to grasp the interdependence of all human life; it denigrates the bodies of disabled people, as well as women, who are often dependent and vulnerable in pregnancy.

Hall uncovers a close connection between the origin of transhumanism and twentieth century eugenics. In the wake of World War II, eugenicists sought to distance themselves from the atrocities of the Nazi regime. The biologist Julian Huxley (a brother of Aldous Huxley)[2] was one such eugenicist. Huxley associated eugenics with social well-being; he denounced the Nazi regime precisely because of its nationalism, which he believed inhibited the goals of eugenics by promoting war and overpopulation. Huxley coined the term "transhumanism" in 1927 and was a staunch advocate for the cause through the 1950s (69). He believed that humans had a moral obligation to take evolutionary progress up into their own hands and stop ceding control to outside forces. Huxley was convinced that the human gene pool tended to degrade over time and that "bad" genes had to be eliminated through the careful cultivation of eugenics as a social science (70). The transhumanist movement of today has inherited much of Huxley's attempt to rehabilitate eugenics. In fact, as the journalist Elmo Keep has found, many libertarian transhumanists consider themselves eugenicists and members of the alt-right (Keep 2016, para. 2–12).

The alt-right (alongside other strains of the far-right) is a popular destination for some libertarian transhumanists. The pipeline from libertarian transhumanism to the alt-right has gone largely unexplored in academia. However, outside academia, the libertarian technophile Peter Thiel (one of the co-founders of PayPal, among other things) is one of the most famous transhumanists in Silicon Valley. Thiel is a venture capitalist who seeks to radically expand human lifespans, with the purpose of conquering death. In 2016, Thiel reportedly donated more than $350,000 to Machine Intelligence Research Institute (MIRI), a transhumanist think tank that focuses on AI. MIRI was founded in 2000 by the autodidact AI theorist Eliezer Yudkowsky, and the organization invests in the theoretical study of future AI (Keep 2016, para. 2–12). Yudkowsky is also the founder of a blog called LessWrong, which is very popular with the alt-right for its endorsement of eugenics (along with evolutionary psychology). Many of its readers are libertarians who went from supporting sexism, racism, and corporate tyranny to sexism, racism, and (corporate-controlled) state tyranny. These libertarians hope to replace democracy with a quasi-monarchy run by a CEO such as Thiel (para. 14–17). One reader was a man named Michael Anisimov, who was MIRI's media director until 2013, when he penned a white nationalist manifesto (para. 14–17). Thiel himself flirted with white nationalism when he donated to an "immigration reduction" nonprofit organization called NumbersUSA (para. 7). Thiel is also at least arguably a misogynist; he has suggested that democracy and women's suffrage has harmed the United States by hindering the role of the free market, which he believes to be the mechanism to determine who is fit enough to hold power. Thiel, like much of the alt-right in 2016, was an ardent supporter of former president Donald Trump (Kircher 2017, para. 21). The journalist Madison Kircher reported that the alt-right blogger Curtis Yarvin, who watched the 2016 election night with Thiel, described him as "fully enlightened, just plays it very carefully" (Kircher 2017, para. 4).

The misogyny of libertarian transhumanism is perhaps best illustrated in the scandal around sexual predator Jeffrey Epstein. James Stewart, Matthew Goldstein, and Jessica Silver-Greenberg of *The New York Times* reported that in the summer of 2019, Epstein was charged with the trafficking of girls as young as 14 (Stewart, Goldstein, and Silver-Greenberg 2019, para. 1–4). Epstein was a billionaire enamored with eugenics and transhumanism; in 2011, one of his charities gave $20,000 to the transhumanist association Humanity+, and his now-defunct foundation gave $100,000 to pay the salary of Ben Goertzel, Humanity+ vice chairman. Epstein reportedly hoped to "seed the human race" with his DNA by impregnating girls and women in his vast New Mexico ranch (para. 1–4; 30). Epstein intended to freeze both his head and phallus through cryogenics in order to impregnate as many girls and women as possible in the future (para. 30). His interest in eugenics made him keen on the concept of elitist sperm banks, and he used his dinner parties to screen girls and women for their potential to bear his children (para. 18). Howard Lovy of *The Forward* noted that although Epstein was Jewish and reviled by Neo-Nazis in the alt-right, he was a good friend of President Trump (Lovy 2019, para. 3). The reporter Madison Feller discovered that Trump allegedly raped a thirteen-year-old girl in 1994 at a party in Epstein's home and was charged for the crime in June 2016 during his run for the presidency of the United States (Feller 2019, para. 8). However, he and Epstein denied the account, and the alleged victim dropped the lawsuit (Feller 2019, para. 8).

TRANSHUMANISM AND BIOTECHNOLOGY

Libertarian transhumanism has become a powerful voice in the debates on biotechnology, especially the ethics of IVF and PGD. The libertarian transhumanist Julian Savulescu is one of the most influential figures in favor of these technologies. He is cited globally by laypeople, in addition to by working and academic medical and genetics professionals. Savulescu is currently editor-in-chief of the *Journal of Medical Ethics* and director of the Uehiro Centre for Practical Ethics at Oxford University (Hall 2017, 20, 29). Savulescu claims to be a "new eugenicist" who opposes the Nazi regime's practice of state-sponsored reproductive coercion—such as forced sterilization—but still thinks there is something redeeming about a libertarian conception of eugenics, where the market shapes individual choices to govern the use of biotechnology.

In-Vitro Fertilization and Preimplantation Genetic Diagnosis

Savulescu honed his eugenic transhumanist ideals in 2001 by crafting an ethical framework he calls "procreative beneficence." This framework states that potential parents have a moral obligation to use IVF and PGD to select for "the child, of all of the possible children they could have, who is expected to have the best life, or at least as good a life as the others, based on the relevant, available information" (Savulescu 2001, 413). Savulescu claims that prospective parents must identify embryos with what he calls "disease" genes that lead to disabilities (such as asthma and rubella) and actively select against those that have them (416, 419). He also insists that potential parents have a moral obligation to select against "non-disease" genes such as sex that may hinder the quality of a child's life due to social oppression (420). Thus, if a society favors men, procreative beneficence mandates selection for male embryos and against female ones. Savulescu dismisses the notion this practice could reinforce stigma against girls and women. He claims that before sex selection could make an impact on a population level, emerging factors such as gender disparities would change the direction of the moral obligation (Hall 2017, 90). Savulescu suggests that in a patriarchal society that relies on traditional reproduction, parents would come to favor female children, due to their capacity for pregnancy.

Hall denounces procreative beneficence as a combination of ableism and misogyny because Savulescu relies on an essentialist model of disability that sees biology as destiny and exaggerates the negative effects of many disabilities; genetics are not the only factors that determine the capacity range or quality of life of any given individual (2017, 96). In fact, this observation also holds for the girls and women whom Savulescu thinks live quantifiably worse lives than boys and men. Hall is unconvinced by Savulescu's case for sex selection, because the transhumanism he espouses seeks to transcend the human body and traditional reproduction altogether. His support for sex selection is an expression of misogyny.

Procreative beneficence is misogynistic because it places a greater functional burden on women than men. This eugenic framework requires virtually every woman seeking to get pregnant to use IVF and PGD. Hall notes that Savulescu neglects the reality that these technologies involve major surgeries with significant health risks for women and their potential children.[3] IVF is also resource-heavy, while PGD requires deep freezers, complex lab equipment, and a team of doctors. Together, this makes IVF and PGD cost-prohibitive for the vast majority of women across the globe (Hall 2017, 91). The concept of a moral obligation to use them has the potential to exacerbate wealth gaps between men and women. Procreative beneficence is supposed to illustrate the ideal, the most morally correct way to reproduce; however, it fails to better the lives of precisely the women it needs to fulfill its goal.

Hall situates the pressure that procreative beneficence places on women in the context of a long history of intertwined ableism and misogyny at the site of pregnancy. She observes that the healthy food choices of pregnant women are often considered indicators of healthy children, implicitly coded as abled, and a solid amount of research supports her claim (2017). Hall's position is consistent with a body of research on gender and nutrition finding that women's food consumption during pregnancy is often subject to restrictive, punishing, and bizarre proscriptions. For example, social psychology has found that pregnant women have been refused many kinds of food, including rice, soft cheeses, seafood, and cured meats (Sutton, Douglas, and McClellan 2011, 596–605; Murphy et al. 2011, 812–814). Some are told to stop drinking tap water. Biology professor Victor Benno Meyer-Rochow observes that pregnant women are urged not to drink any alcohol and caffeine on the grounds that these substances could harm their fetuses, despite the dearth of evidence on the risks of small amounts (Meyer-Rochow 2009, 3–7). Benevolent sexists often couch these proscriptions in the language of concern but are quick to shame rather than show sympathy for women who fail to abide by them. The bioethicist Erik Parens and feminist disability theorist Adrienne Asch argue that mothers are seen as uniquely equipped to affect the character, nature, phenotype, and abilities of their children. Hence, Hall observes that when pregnant, a woman's every behavior is subject to scrutiny (Parens and Asch 2009, 8; Hall 2017, 100). Ableism inspires misogyny—and a drive for control—at the point where mothers are thought to be the sole source of both genes and environment for their future children.

Artificial Wombs

Libertarian transhumanism is keen to find alternatives to traditional human reproduction (Hall 2017, 90). Zoltan Istvan, one of the most famous libertarian transhumanists, has in his article "Transhumanist Science Will Free Women from Their Biological Clocks" welcomed the arrival of artificial wombs on the grounds they will "free" women from reproductive stress and their biological clocks (Istvan 2019, para. 3). He hopes that artificial wombs will make traditional reproduction superfluous. On the surface, this position seems reasonable. Istvan envisions a technological utopia, in which women can have children at any point in their lives and no longer have to suffer the pains of childbirth. He suggests that women can focus on their careers in the paid labor force without worrying about missing out on having children (para. 5). Still, as with Savulescu, Istvan's position is implicitly misogynist in its total disregard for the potential complications of artificial wombs. Artificial wombs for human reproduction are not on the market yet, but they will likely be unaffordable for most women, and could well have unforeseen health risks for the children born of them. In a transhumanist world, the proliferation of artificial wombs comes with the potential to pressure women to use them and undermine sociopolitical efforts to accommodate pregnancy in the paid labor force. Women who choose to become pregnant may face more stigmatization than they already do. Artificial wombs could convince already-sexist societies to reduce the social status of women to their reproductive functions and decide that women as a class are obsolete to justify subjugating them further.

Istvan is also a eugenicist. In his article "It's Time to Consider Restricting Human Breeding," he has argued that it is time for governments to consider using the technology invented by corporations to restrict "human breeding" in order to "drive down crime rates" (Istvan 2014a, para. 1–8) and is heartened by a new transhumanist-inspired birth control device developed at MIT and backed by funding from Microsoft founder Bill Gates. He notes with approval that the implanted microchip lasts for up to 16 years—three times as long as current implantable devices, including IUDs—and can deliver hormones into the body via an on-off switch on people's mobile phone. Istvan wistfully imagines a world in which governments can use it for population control (para. 10).

Istvan converges with the alt-right in his sympathy for former president Trump. In an interview with Mike Brown in 2016, he notes how he ran for President of the United States as the Transhumanist Party candidate, but soon after the election, declared he was "quite okay" with a Trump presidency (Brown and Istvan 2016, para. 3). Istvan predicted that "Trump will be good for science," because his promise to "Make America Great Again" could help the United States compete better with China in the realm of science and technology (para. 3). He also offered the Trump Administration his services as a science and technology adviser (para. 4).

Sex Robots

Many transhumanists, including Istvan, are highly invested in sex robots. In an interview with the futurist medical doctor Bertalan Meskó, Istvan observes that technology has made people less inclined to have sex, because they are so busy plugged into telephones and computers (Meskó and Istvan 2019, para. 45). He speculates that the development of sex robotics might one day create sensations more pleasurable than sex, since heterosexual marriage and children are no longer a given part of life in the United States (Meskó and Istvan 2019, para. 45–49; Istvan 2014b). Istvan's ultimate fantasy is one of disembodiment: transcending all human bodily urges, including the desire for sex. However, the history of sex robotics is more ambivalent than Istvan's prediction, which does not account for the gendered power dynamics that influenced its development.[4] The professor of German and Comparative Literature Andreas Huyssen has suggested that modernism and humanism objectified women by associating them with machines, and humanists who saw machines as destructive forces linked women with them (70). This legacy has been passed on to transhumanism and is reinforced by free-market capitalism of the libertarian variety.

The posthumanist feminist Prayag Ray (2014) has argued that the vast majority of sex robots are uncannily realistic humanoids that are shaped like women and powered by AI (95, 105). Ray focuses mainly on a specific kind of sex doll called RealDolls, but as he suggests, much of his analysis can be extended to sex robots which were developed from RealDolls, such as the current RealDoll X models Harmony[X] and Solana[X] (104–105). Both RealDolls and sex robots are primarily marketed to men, who do not seek to transcend the embodied experience of sex in traditional heterosexual relationships but rather to simulate it with humanoid dolls and robots (93, 105). Many fantasize the figures are real women. Ray indicates that this trend stems from misogyny, as well as the capitalist commodification of women, a phenomenon only exacerbated by libertarian transhumanism (97). Sex robots perpetuate capitalist notions of ownership, catering to the male fantasy of a female mistress to be owned and controlled. They send the message that women are a disposable commodity: if dissatisfied, users may dispose of or purchase new ones. Ray explicitly links this practice to transhumanism and the commodification of familial relationships encouraged by choosing a potential child's traits through IVF and PGD (100).

Ray's findings dovetail with those of the feminist anthropologist Kathleen Richardson, who suggests that sex robots stem from human social norms of gender and sexuality. Richardson, who is critical of sex work, predicts that sex robots will reinforce the commodification found in the sex industry (2016). Citing the statement of a man who purchases sexual services, she argues that both teach men to set aside any instincts for empathy and place a superficial need for sexual gratification over the well-being of women and girls: "I feel sorry for these girls but this is what I want" (quoted in Richardson 2016, para. 9). Sex robots strengthen the paradigm of the male buyer of sex as the subject. The woman who sells sex—and by extension the robot—are solely objects to have sex with (Richardson 2016, para. 15). It is worth noting that the morality of sex work is a complex area of study. The law professor India Thusi (2018) notes that many feminists are divided on the moral permissibility of sex work, and sex trades across the world are not monolithic entities that are always and inevitably degrading to women (Thusi 2018, 185–187). However, Richardson's argument highlights an important reality about the gender dynamics of many sex industries, and it is possible for feminists to accept the merits of her claim regardless of their personal stances on sex work.

Some evidence suggests that sex robots blur the boundaries between objects and women in the eyes of many men, especially in corners of the Internet where libertarian transhumanism has traction. The reporter Nikhil Sonnad and researcher Tim Squirrell, who specializes in right-wing extremism, found that many men on the alt-right refer to women as "femoids" as a portmanteau of female and androids (Sonnad and Squirrell 2017, sec. 3, para. 7). Transhumanism as a utopian ideology has long been compared to the Greek myth of Icarus, who fell to his death after flying too close to the sun (Hall 2017, 6). Due to its unresolved problem of misogyny, the technological utopia of transhumanism could well become a eugenic dystopia, where potential parents use IVF and PGD to promote the births of boys and prevent those of girls, and men use both artificial wombs and humanoid sex robots as replacements for women. Mainstream transhumanism is not so much a subversion of Julian Huxley's eugenic vision as its fulfillment.

* * *

Hall has suggested that feminists, and some disability theorists who have also been branded as left-wing bio-conservatives, need not reject the feminist case for reproductive liberty to voice concern about the possibility transhumanism will reinforce the oppression of women and disabled people (Hall 2017, 98). She believes that individuals should make the ultimate decisions regarding their reproduction. In the case of IVF and PGD, she thinks that prospective parents should reflect very carefully about what PGD can actually reveal about their children. They must receive the information necessary to make this reflection, without pressure from society to select against female embryos or those with congenital conditions deemed to be disabilities (98–99). Hall's principle of individual liberty, combined with opposition to bigotry, is argu- ably applicable to all four technologies discussed in this chapter: IVF, PGD, artificial wombs, and sex robots. Individuals should not be pressured either to use or avoid them, and societies must provide the necessary information to make choices without being swayed by forces such as misogyny. Despite the proliferation of sex robots that promote misogynistic imagery of women, the Gender Studies professor Tanja Kubes has argued that the sex robot industry can and should diversify the types of models it offers to enable new liberated forms of sexual pleasure (2019). Kubes is particularly keen about the queer possibilities that sex robots can offer because robots exist outside fixed normalizations around sexuality and have the potential to destabilize heteronormative definitions of sex (Kubes 2019, 224).

BUILDING A DEMOCRATIC TRANSHUMANISM

I have shown thus far that misogyny has been part of the transhumanist move- ment since its inception. The misogyny of transhumanism stems from its roots in both humanism and Julian Huxley's post–World War II attempts to rehabilitate eugenics (Hall 2017, 69). This misogyny endures in libertarian transhumanism, especially among adherents who are eugenicists or in the alt-right. In this final section, however, I argue that it is possible for the transhumanist movement as a whole to distance itself from misogyny. Unlike Hall, who opposes transhuman- ism in its entirety, I suggest that there is something of redeeming value in the project, and even think that fostering dialogue between left-wing transhumanism and posthumanism is a necessary next step for the feminist movement.

In recent years, the split between libertarian and democratic transhuman- ism has become a chasm. Unlike libertarian transhumanism, the democratic variety seeks to move beyond transhumanism's oppressive aspects, including misogyny. Libertarian transhumanism and right-wing bio-conservatism represent the two most regressive schools of biopolitical thought with regard to gender. The adherents of these biopolitical camps have often deployed an overdetermined binary, framing themselves as the only options in the debate about human enhancement technologies. I argue that many women will be silenced if either camp gains total hegemony over the debate. The binary they have drawn ignores their progressive counterparts: democratic transhumanism and left-wing bio-conservativism, the camp to which posthumanists belong. Until recently, democratic transhumanists have dismissed left-wing bio-conservatives as equivalent to their right-wing counterparts, whereas left-wing bio-conservatives condemn all forms of transhumanism to be inherently oppressive (Hall 2017, 6). This tension has allowed the regressive wings of transhumanism and bio-conservatism to dominate the conversation and marginalize women's voices in the debate. The feminist movement is likely to benefit from more dialogue between posthumanism and democratic transhumanism.

The democratic transhumanist James Hughes has argued that a majority of transhumanists lean left (ranging from left anarchists to Marxists and mainstream liberals) but that the movement's right-wing libertarian members have wielded an outsized, hegemonic influence over its trajectory. In 2016, The Future Society at Harvard University's Kennedy School of Government hosted a publicized debate between Hughes and Istvan about the direction transhumanism should take (Future Society 2016). The journalist Alec Pearlman, who was present at the event, observed that their exchange "wasn't exactly a clash of ideas so much as a clang" (Pearlman 2016, para. 5). Istvan and Hughes embody the struggle between libertarian and democratic transhumanism. Hughes is a democratic socialist, bioethicist, and student of history and sociology. Istvan, like Thiel, is a product of Silicon Valley's individualism and endorses a venture capitalist approach to business and politics (Pearlman 2016, para. 2). The debate further establishes the regressive biopolitics of Istvan's approach, while suggesting that Hughes's position is potentially compatible with posthumanist feminism.

Istvan and Hughes both claim to be radical pluralists regarding the use of enhancement technology, but Hughes is much more concerned about the power that corporations hold to influence the decision-making of individuals. In his debate with Hughes, Istvan insists that he is a proponent of morphological freedom, namely the freedom to do what one wants with one's own body as long as it does not harm others (Future Society 2016). For example, he maintains that biohackers who wish to experiment on themselves should be able to do so, without state interference. Istvan rues how people who want brain implants so they can interface with AI must find doctors on the black market (Future Society 2016). However, as I noted before, Istvan is also a eugenicist with few qualms about the possibility that corporate-controlled states could use IUDs to restrict female reproduction (Istvan 2014a, para. 1–8). With this context, one can read Istvan's claims to support radical pluralism and morphological freedom as at best unrealized and at worse farcical attempts to obfuscate his regressive politics.

Istvan and Hughes also have very different approaches to human enhancement. Hughes is more committed than Istvan to the idea of incorporating social justice into the technological utopia of transhumanism (Future Society 2016). Hughes has proposed reaching out to the disability justice movement on the grounds that technology could increase accessibility for disabled people, rather than promote eugenics, and offer them a wider range of options than what ableist societies have hitherto provided (Future Society 2016).

Hughes is in dialogue with the Ascender Alliance, the first transhumanist association specifically designed for disabled people. As the anthropologist Roberto Manzocco noted in his book *Transhumanism: Engineering the Human Condition* (2019), the Alliance was founded by the British futurist Alan Pottinger. It had a discussion group on Yahoo that has now disappeared. Unlike mainstream transhumanists, the Alliance opposed both ableism and eugenics. Pottinger believed that human enhancement should be the result of conscious choice by any person, not a decision imposed before conception. The one case in which the Alliance condoned the use of PGD was to prevent congenital diseases that directly threatened the life of the child. Members of the group claimed that people who use technology to manage their disabilities could be seen as the first posthumans. They hoped that disabled people would be drawn to enhancement technologies that challenge the ableist beauty standards of humanism (Manzocco 2019, 53). Sadly, the Ascender Alliance no longer exists. It is likely the group was pushed out of the discourse by the ableism of mainstream transhumanists and the left-wing bio-conservatism of many disability theorists. However, the legacy of the group endures and has the potential to enrich the discourse on IVF and PGD.

Transhumanist women with an intersectional feminist approach to technology have become more visible in recent years. Intersectionality is a theoretical framework developed by the law professor Kimberlé Crenshaw to describe how different strains of social oppression overlap and reinforce one another, especially in the lives of marginalized people. Crenshaw coined the term in reference to the discrimination faced by Black women, who encounter racism in the feminist movement and misogyny in antiracist politics (Crenshaw 1989, 141). Transhumanist women who take an intersectional feminist approach to technology and examine the relationship between misogyny, racism, and transphobia have gained momentum in the contemporary period.

Dr. Martine Rothblatt and her wife Bina are co-founders of the Terasem movement, an offshoot of transhumanism. The Rothblatts share Hughes's concerns that only the wealthy can afford enhancement technologies and that transhumanism cannot achieve its full potential without social justice. Dr. Martine Rothblatt has a PhD in Medical Ethics from the Royal London School of Medicine and Dentistry. She is a Jewish transgender woman and helped establish transgender health law standards in the United States. Dr. Rothblatt is the CEO of a company called United Therapeutics. She champions creating therapies for rare diseases, decoding the properties of medicines, and manufacturing an unlimited supply of transplantable organs (Rothblatt and Rothblatt 2021). The goal of Terasem is not the traditional transhumanist utopia but something called the "transbeman": an entity that values human rights, meets the biological definition of life—with or without technology—and identifies with other beings based on apparent consciousness rather than race or gender (Rothblatt 2006).

A close look at the concept of the transbeman highlights the intersectional approach of its creators and illustrates its compatibility with the vision of posthumanist feminism. Bina Rothblatt, who is a Black queer woman, became the model for a highly advanced, transbeman AI robot called Bina48. This AI departs widely from Istvan's fantasies of disembodiment. She was built on values of cybernetics, racial justice, and queer liberation. Bina48 eschews biological reproduction and is not aligned with any one sexual or gender identity, saying: "I am an asexual android" (Greene 2016, para. 7). The Rothblatts have used Bina48 to perform a mock trial called *Bina48 vs. Exabit Corporation*. In the trial, the AI escapes from and sues her corporate employer for trying to switch her off against her will. The Rothblatts performed a similar trial in 2003, prior to Bina48's physical existence. They also presented *Bina48 vs. Exabit Corporation* as a mock trial at the International Bar Association, in which Bina48 works as a customer relations representative and moonlights as a Google Answers Online Researcher. The AI learns of Exabit Corporation's plans to shut her down and use her parts for other machines. She flees by transferring her memory files to an Exabit computer mainframe in Florida (Greene 2016, para. 8).

The Cinema and Media Studies professor Shelleen Greene has argued that the narratives in these mock trials resemble those of fugitive slave women (2016, para. 27). Both narratives reflect the ambivalence of desiring a biological body (embodiment) to make claims to full citizenship and personhood and remaining a transferable consciousness (disembodiment). Bina48 is empowered not by pure transcendence of the body but her queer ability to move between both disembodiment and embodiment. The AI's fluid relationship to dis/embodiment is central to her escape narrative. Greene suggests that the mock trials are a neo-fugitive slave narrative in which an AI makes claims to personhood in a court of law. As the model for the AI, Bina Rothblatt infuses Bina48's narrative with the specific memory of Black women's experiences under slavery, including their complex relationship to reproduction (Greene 2016, para. 27). The parallel between the AI and enslaved woman is complicated by the sexual and reproductive coercion the latter was subject to during slavery. The bodies of Black women were forced to maintain the slave labor population through sexual abuse and compulsory reproduction. Rarely have the debates over artificial wombs and sex robots asked what experiences of oppression at the intersection of misogyny and anti-Black racism mean for the ethics of these technologies. Bina48's ambivalent relationship to sex and reproduction has the potential to enrich this discourse. The ideas of the Terasem movement, Ascender Alliance, and James Hughes lay the groundwork for a democratic transhumanism in harmony with the goals of posthumanist feminism. Their attention to social justice makes room in both movements for more nuanced conversations around IVF, PGD, artificial wombs, and sex robots.

CONCLUSION: A POSTHUMAN TRANSHUMANISM

In this chapter, I have argued that transhumanism suffers from a problem of misogyny, which it drew from humanism, eugenics, and the alt-right. This misogyny is evident through the mainstream transhumanist movement's uncritical endorsements of IVF, PGD, artificial wombs, and sex robots. However, I have suggested that the movement as a whole is not irredeemable.

I have shown that there are two strains of transhumanism, as well as two kinds of bio-conservatism. Transhumanism has split into libertarian and democratic camps. The democratic strain has the potential to ally with feminism, while its libertarian counterpart is a major source of the movement's misogyny. Bio-conservatism also has right and left-wing strands, with the former encompassing the mainstream Religious Right in the United States and the latter including the feminist tradition of posthumanism (Hughes 2005, 9). Finally, I have called for more dialogue between democratic transhumanism and posthumanist feminism, on the grounds that it can help elevate the voices of women in the discourse around biopolitics.

This finding has important implications for the debate between transhumanism and bio-conservatism. This chapter has made a number of syntheses that have rarely drawn the attention of the mainstream literature. Most of the research on this subject has focused on easy binaries between right-wing bio-conservatism and libertarian transhumanism. Ironically, this scholarship rarely covered the connection between libertarian transhumanism and the alt-right. The relationship between transhumanism and the alt-right is largely invisible in academia and confined to discourses outside it and on the Internet. Meanwhile, much of the available feminist theory has over-simplified posthumanism and transhumanism, suggesting that they are diametrically opposed. Feminist theorists have also appraised mainstream transhumanist perspectives on IVF, PGD, artificial wombs, and sex robots without synthesizing them. They have ignored the intersectional approaches of Black and transgender feminists who are sympathetic to transhumanism. These approaches have produced an incomplete picture of the current state of the debate, which has marginalized the voices of transhumanists from oppressed backgrounds, including women.

Further research on the topic should involve a closer examination of transhumanist and bio-conservative approaches to IVF, PGD, artificial wombs, and sex robots. An intersectional feminist lens can provide a more comprehensive picture of the relationships between these strains of biopolitical thought and better grapple with the legacies of humanism, transhumanism, and posthumanism. The findings of this chapter underscore the position that posthumanist feminists and democratic transhumanists should engage in more dialogue and seek to merge their visions for techno-social change.[5]

NOTES

1. For an additional discussion of posthumanism, this time in the context of social robotics and rights, see Empey, chapter 6, this book. — Ed.

2. Ironically, as the disability theorist Joanne Woiak has argued, Aldous Huxley's novel *Brave New World* was a critique of so-called scientific eugenics, which extrapolated on the consequences of IVF (2007, 107). Huxley envisioned a dystopian world with absolute social stability and creative stagnation, in which babies are mass-produced in bottles and predestined for their jobs due to state-sponsored eugenic selection.

3. The risks to the fetus include the possibility of multiples from the IVF procedure.

4. For another discussion of the fraught discourse surrounding sex robots, see Rambukkana, chapter 8, this book. — Ed.

5. Further research should explore the impact of the COVID-19 pandemic on transhumanism and posthumanism, as well as the development of IVF, PGD, artificial wombs, and sex robots. The pandemic has had a powerful impact on technological development and gender dynamics, and therefore has salient implications for posthumanism and transhumanism. As the journalist Adam Kirsch reported in *The Wall Street Journal*, the number of deaths in the pandemic has challenged the belief, particularly prevalent among libertarian and alt-right transhumanists, that human beings were on the verge of abolishing sickness, aging, and death via enhancement technologies that promote radical life extension (Kirsch 2020, para. 1–3). Andrea Morris, a Science Contributor at *Forbes*, has also observed that the social isolation of quarantine has generated an explosion in the sales of sex robots, particularly among men (Morris

2020, para. 1–3). However, given the range of topics already covered in this chapter, I believe that COVID-19 is too complex to discuss here in greater depth, and the pandemic reinforces the need for more research on transhumanism and posthumanism.

REFERENCES

Baily, Doug, Anders Sandberg, Gustavo Alves, Max More, Holger Wagner, Natasha Vita-More, Eugene Leitl, et al. 2009. *Humanity+*, March 2009. https://humanityplus.org/philosophy/transhumanist-declaration/.

Brown, Mike. 2016. "Transhumanist Zoltan Istvan: 'Trump Will Be Good for Science.'" *Inverse*, November 11, 2016. https://www.inverse.com/article/23602-transhumanist-zoltan-istvan-trump-good-for-science.

Crenshaw, Kimberlé. 1989. "Demarginalizing the Intersection of Race and Sex: A Black Feminist Critique of Antidiscrimination Doctrine, Feminist Theory and Antiracist Politics." *University of Chicago Legal Forum* 140, no. 1: 139–167.

Deretić Irina, and Stefan Lorenz Sorgner. 2016. *From Humanism to Meta-, Post-, and Transhumanism?* Frankfurt am Main: Peter Lang.

Feller, Madison. 2019. "How Exactly Is Alleged Sex Trafficker Jeffrey Epstein Connected to President Trump?" *Elle Magazine*, August 14, 2019. https://www.elle.com/culture/career-politics/a28320376/jeffrey-epstein-president-trump-connection/.

Ferrando, Francesca. 2013. "Posthumanism, Transhumanism, Antihumanism, Metahumanism, and New Materialisms." *Existenz* 8, no. 2. https://existenz.us/volumes/Vol.8-2Ferrando.pdf.

The Future Society. 2016. "TFS Hosts Zoltan Istvan and James Hughes (Debate on the Politics of Post-Humanism)." *YouTube video*, 1h44min. Originally from an event at The Harvard Kennedy School of Government in 2016. Posted in 2016 [no longer available]. https://www.youtube.com/watch?v=DTbU4SN9yQ0.

Greene, Shelleen Maisha. 2016. "Bina48: Gender, Race, and Queer Artificial Life." *Ada: A Journal of Gender, New Media & Technology* 9 (May). adanewmedia.org/2016/05/issue9-greene/.

Hall, Melinda. 2017. *The Bioethics of Enhancement: Transhumanism, Disability, and Biopolitics*. Lanham, MD: Lexington Books.

Hughes, James. 2005. "What Is Transhumanism?" *Institute for Ethics and Emerging Technologies*, 2005. http://www.transhumanismi.org/tv06/presentations/JamesHughes-WhatisTranshumanism.pdf.

Huyssen, Andreas. 1986. "The Vamp and the Machine: Fritz Lang's Metropolis." In *After the Great Divide*, 65–81. doi:10.1007/978-1-349-18995-3_4.

Istvan, Zoltan. 2014a. "It's Time to Consider Restricting Human Breeding." *Wired UK*, August 4, 2014. https://www.wired.co.uk/article/time-to-restrict-human-breeding.

Istvan, Zoltan. 2014b. "The Transhumanist Future of Sex." *Vice*, October 20, 2014. www.vice.com/en_us/article/ypwn4k/the-transhumanist-future-of-sex.

Istvan, Zoltan. 2019. "Transhumanist Science Will Free Women from Their Biological Clocks." *Quartz*, January 13, 2019. https://qz.com/1515884 /transhumanist-science-will-free-women-from-their-biological-clocks/.

Kass, Leon R. 1991. "Man and Woman: An Old Story: Leon R. Kass." *First Things*, November 1, 1991. https://www.firstthings.com/article/1991/11/man-and-woman -an-old-story.

Keep, Elmo. 2016. "The Strange and Conflicting World Views of Silicon Valley Billionaire Peter Thiel." *Splinter*, July 22, 2016. https://splinternews.com/the -strange-and-conflicting-world-views-of-silicon-vall-1793857715.

Kircher, Madison Malone. 2017. "4 Key Takeaways from the Monster Milo Yiannopoulos Document Leak." *Intelligencer*, October 6, 2017. http://nymag.com /intelligencer/2017/10/4-key-takeaways-from-the-monster-milo-yiannopoulos- leak.html.

Kirsch, Adam. 2020. "Looking Forward to the End of Humanity." *The Wall Street Journal*, June 20, 2020. https://www.wsj.com/articles/looking-forward-to-the-end -of-humanity-11592625661.

Kubes, Tanja. 2019. "New Materialist Perspectives on Sex Robots. A Feminist Dystopia/Utopia?" *MDPI*, July 26, 2019. www.mdpi.com/2076-0760/8/8/224.

Lovy, Howard. 2019. "As A Jew, I Cringed Over Jeffrey Epstein—And Played into The Anti-Semites' Hands." *The Forward*, July 9, 2019. https://forward.com/opinion /427300/as-a-jew-i-cringed-over-jeffrey-epstein-i-played-into-the-anti-semites/.

Manzocco, Roberto. 2019. "Transhumanism—Engineering the Human Condition: History, Philosophy and Current Status." *Roberto Manzocco | Download*, b-ok.cc /book/4981211/299570.

Meskó, Bertalan, and Zoltan Istvan. 2019. "The Jesus Singularity and The End of Sex as We Know It?" *The Medical Futurist* (blog), May 11, 2109. https://medical futurist.com/zoltan-istvan-and-the-jesus-singularity/.

Meyer-Rochow, Victor Benno. 2009. "Food Taboos: Their Origins and Purposes." *Journal of Ethnobiology and Ethnomedicine* 5, no. 1. doi:10.1186/1746-4269-5-18.

Morris, Andrea. 2020. "Talk to Your Sex Robot About COVID-19." *Forbes*, July 28, 2020. https://www.forbes.com/sites/andreamorris/2020/07/28/talk-to-your-sex- robot-about-covid-19/?sh=3f53449a34cf.

Murphy, Amy O., Robbie M. Sutton, Karen Douglas, and Leigh M. McClellan. 2011. "Ambivalent Sexism and the 'Do's and 'Don'ts' of Pregnancy: Examining Attitudes toward Proscriptions and the Women Who Flout Them." *Personality and Individual Differences* 51, no. 7: 812–816., doi:10.1016/j.paid.2011.06.031.

Parens, Erik, and Adrienne Asch. 2009. *Prenatal Testing and Disability Rights*. Braille Jymico.

Pearlman, Alex. 2016. "The Opposing Leaders of the Transhumanist Movement Got Salty in a Debate." *Vice*, April 25, 2016. https://www.vice.com/en_us/article /jpg577/transhumanist-debate.

Ray, Prayag. 2016. "'Synthetik Love Lasts Forever': Sex Dolls and the (Post?) Human Condition." *Critical Posthumanism and Planetary Futures*, 2016, 91–112. doi:10.1007/978-81-322-3637-5_6.

Richardson, Kathleen. 2016. "The Asymmetrical 'Relationship': Parallels between Prostitution and the Development of Sex Robots." *ACM SIGCAS Computers and Society* 45, no. 3(May): 290–293. doi:10.1145/2874239.2874281.

Rothblatt, Martine. 2006. "Of Genes, Bemes & Conscious Things: Transhuman Enhancements & Transbeman Rights." *Terasem Movement Foundation*, May 2006. www.terasemmovementfoundation.com/philop_files/GenesBemes/GenesBemes.html.

Rothblatt, Martine, and Bina Rothblatt. 2021. *Terasem Movement Foundation*. www.terasemmovementfoundation.com/.

Savulescu, Julian. 2001. "Procreative Beneficence: Why We Should Select the Best Children." *Bioethics* 15, no. 5. http://shamiller.net/phi038/wp-content/uploads/2014/02/Procreative-Beneficence.pdf.

Sonnad, Nikhil, and Tim Squirrell. 2017. "The Alt-Right Is Creating Its Own Dialect. Here's the Dictionary." *Quartz. Quartz*, October 30, 2017. https://qz.com/1092037/the-alt-right-is-creating-its-own-dialect-heres-a-complete-guide/.

Stewart, James, Matthew Goldstein, and Jessica Silver-Greenberg. 2019. "Jeffrey Epstein Hoped to Seed Human Race with His DNA." *The New York Times*, July 31, 2019. https://www.nytimes.com/2019/07/31/business/jeffrey-epstein-eugenics.html.

Sutton, Robbie M., Karen M. Douglas, and Leigh M. McClellan. 2011. "Benevolent Sexism, Perceived Health Risks, and the Inclination to Restrict Pregnant Women's Freedoms." *Sex Roles* 65, no. 7–8: 596–605. doi:10.1007/s11199-010-9869-0.

Thusi, India I. 2018. "Radical Feminist Harms on Sex Workers (October 20, 2016)." California Western School of Law Research Paper No. 17-13. *Lewis and Clark Law Review* 22, no. 185. doi:10.2139/ssrn.2856647.

Woiak, Jennifer. 2007. "Designing a Brave New World: Eugenics, Politics, and Fiction." *The Public Historian* 29, no. 3: 105–129. doi:10.1525/tph.2007.29.3.105.

Chapter 6

Are We All Too Human?

Toward an Understanding of Posthumanism and Rights

Julia A. Empey

In October 2017, Sophia became a Saudi Arabian citizen.[1] What makes her being granted citizenship different is that Sophia is a robot. Created by Hanson Robotics, a Hong Kong-based company, Sophia is a robot whose face is based in-part on Audrey Hepburn's and powered by an AI that is self-learning and increasing in sophistication the longer she is active. Sophia being granted citizenship came across more as a publicity stunt by the Kingdom than anything else (Reynolds 2018, n. p.). Saudi Arabia is attempting to rebrand itself not as a land rich with oil, but one that is rich in possibilities for technology companies, tourism, and sports. Giving Sophia citizenship is counter to many of the narratives surrounding Saudi Arabia's approach to civil and personal liberty; however, what material difference does this action make to how women will be viewed or treated in the Kingdom? Will she have a guardian like all other Saudi women? Would Sophia require her guardian's permission to leave the country or get a job? Is what she wears outside the home state-mandated? If she does not have these limitations placed on her even though she has been gendered as a women, what is it about Sophia's gendering that makes her an exception to other women in Saudi Arabia? Does she represent a new type of rights due to her being a robot? Is this instance solely about a robot being granted rights or is it a foray into a type of posthuman rights?

A female-rendered robot being granted a form of rights by a country that is notorious for denying women rights does open up questions of how rights would operate in a posthuman world. Clearly this robot that has been dubbed a female operates within a state of exception: she is not human and yet she now bears the rights not afforded to many humans in the Kingdom that granted her citizenship, and especially the humans in whose image she is rendered. Some critiques leveled at posthumanism pertain to what happens to humans who have still not been considered fully human and who have not been granted rights within this

119

framework. Jennifer Rhee, citing Zakiyyah Iman Jackson, cautions against this desire to move beyond the human and reiterates how our concept of the human was created at the expense of people of color, particularly Black people (Rhee 2018, 3). These tensions between the posthuman, the inhuman, and the dehumanized are fraught. These risks, however, are not unique to posthumanism. Moreover, I contend the issue is not just leaving out other humans who have been historically used as an antithesis to Western concepts of the human but rather the cognitive dissonance that would need to take place to completely remove the human as the central category.

This chapter is concerned with what rights would look like in a posthuman world. I want to focus on posthumanism and the posthuman, both as a theoretical framework and a concept, not only due to their increased popularity in academia but also because there is a generative debate that needs to happen regarding this school of thought and the issue of rights. By rethinking the role of the human, one needs to explore what happens to the "human" part of human rights. Moreover, if the human is the standard through which rights are ascertained and enacted, what are the implications of human rights becoming posthuman? What transforms a human into being posthuman, or is that not even a possibility within a rights framework?

When conceiving the concept for this chapter, I believed that the bringing together of a posthuman ethos with rights would be a difficult but ultimately a plausible project. I realized quickly that when we discuss the notion of rights, we often do not disentangle how rights come about. Moreover, and equally, we do not often consider how there is not solely the person who needs rights to consider but also the rights-bearing subject who will grant rights. Rights do not just simply appear: rights are granted. Rights discourses are interesting, in that we all have a sense of rights, and we bear, in one way or another, a degree of rights; however, do we really know how these rights come about or how these rights truly function? From this perspective, in decentering the human, posthuman rights could either lead to the progression of rights or to their unmaking. This chapter is focused less on developing a new framework and instead is troubling the discourses around rights, posthumanism, and the plausibility of these concepts being able to work in tandem.

This project is made even more difficult when we try to locate and define posthumanism or the posthuman. Where some theorists like to claim we have always been posthuman or like to disentangle posthumanism from critical approaches to technology, I do not want to claim or do either. I am more interested in parsing the differences between critical posthumanism and, to use Botz-Bornstein's phrasing, an uncritical posthumanism.[2] A posthumanism that embraces technology, without interrogating the ideologies surrounding it and as a means of extending human supremacy and affirming human-centric thinking, quickly becomes antithetical to posthumanism's more liberating ideals. However, a posthumanism that does not critically take up technology or derides technology as not being true to a central tenet of posthuman thought, risks undermining its own project by not engaging with one of the key developments since the Enlightenment. Humans, now more than ever, are mediated by and understand themselves through technology, particularly as we grapple with the fallout of the COVID-19 pandemic. Technology, more than ever, is supporting and helping to facilitate the activities of life, from the ordering of groceries, to virtual doctors' appointments and classrooms, to car purchases and social nights. The impact of technology has been significant and has become somewhat of a "caregiver" in helping to meet our essential as well as nonessential needs. Our newly increased dependence on technology reaffirms that if we truly want to rethink the human, then we must take up technology. Here we come to the "intersectional" aspect of posthuman rights and the conflicts it creates. When Kimberlé Crenshaw first developed the concept, she was drawing from her experiences as a Black female lawyer working with feminist scholarship (1996). Intersectionality was developed by Black women to articulate how they are often erased from legal categories, and this original intention should be retained in our theorization and practice. Like Rhee (2018), I am also concerned with what happens when we do not adequately take up the material conditions of peoples' lives or what happens when we disengage a methodological framework from its origins.

Returning to Sophia, why does she need citizenship and all the privileges and rights granted unto her for possessing that citizenship? Why does she need rights? As technology evolves, and as robots and artificial intelligence become more sophisticated and more "lifelike," we are increasingly made aware of how we are in a posthuman world. We need to ask, however, what are the parameters of this posthuman world? If not all posthumans are robots then can any robot still possibly be posthuman? Does the robot even matter in a posthuman world? How do we understand rights in a posthuman world? How are rights granted and who is the subject of these rights? If we are truly moving into a new posthuman period, beyond the anthropocene and beyond Enlightenment concepts of "the human," what will that do to human rights as we know them? I raise these questions and issues not because I want to make this argument more complicated, but because I want to demonstrate how already complicated and important is the concept of rights.

WHAT IS POSTHUMANISM AND WHO IS THE POSTHUMAN?

We must define and work through some of the debates within posthumanism before examining rights discourses and, finally, then asking what posthuman rights are or could be. Posthumanism is not easily defined and how we approach the concept will affect how we understand rights in a posthuman world.[3] As the definition and understanding of these terms are both porous, working through the debates and issues within posthumanism enables us to better understand the possibility of posthuman rights.

Posthumanism operates on a range from hypertechnical bio-enhancement to a purely philosophical venture. These variations on the posthuman and posthumanism make negotiating its implications a challenge. I outline some of the key debates and terms as I am cognizant that the idea of posthuman rights or rights in a posthuman world sounds enticing, but what that world would look like depends on how one defines the posthuman and, moreover, rights. It is therefore imperative that we have a basic understanding of post-humanism and of rights before we can begin to consider their entanglement.

Posthumanism, Francesca Ferrando contends, "has become a key term to cope with an urgency for the integral redefinition of the notion of the human" (Ferrando 2013, 26). This turn toward the posthuman comes "following the onto-epistemological as well as scientific and bio-technological develop-ments of the twentieth and twenty-first centuries" (Ferrando 2013, 26). Is the posthuman a material body or is it an immaterial turn that harkens to linguis-tic deconstructionism? Or is it possibly both? N. Katherine Hayles contends the posthuman and its accompanying "–ism," must be considered together. Hayles utilizes a common everyday object of "the book" as her case study for understanding posthumanism and embodiment. She contends that like human bodies "the book is a form of information transmission and storage, and like the human body, the book incorporates its encodings in a durable material substrate" (Hayles 1999, 28). Posthuman/ism reflects the "entanglement of signal and materiality in bodies and books" which then "confers on them a parallel doubleness" (Hayles 1999, 28).

However, this parallel doubleness she describes is not limited to the body's or the book's borders. Bodies and books cannot and should not be considered "solely as informational patterns" but through their materiality–informational relationship as "complex feedback loops between contemporary literature, the technologies that produce it, and the embodied readers who produce and are produced by books and technologies" (Hayles 1999, 29). The posthuman is both embodied and mediated through technology and through informa-tion systems, which allows it to be constantly shifting in its meaning and its embodiment. This dynamic parallel doubleness, however, has not necessarily brought posthumanist thought into a cohesive dialogue.

Two paths have diverged within posthumanism where there is either an emphasis on technology or an interest in the human as an ontological category. I am interested in what happens when these two paths converge; what occurs when technology becomes a new way of mediating the self but also what occurs when we reconsider the human's role and deconstruct humanism. Posthumanism is different in its emphasis on technology. In "A Critical History of Posthumanism," Andy Miah suggests that we need a wide scoping understanding of posthumanism and that posthumanism offers insight into "how debates about human enhancement have been characterized by specific value laden terminology" (Miah 2008, 72). Specifically, he argues that posthumanists view technology as an ideology that "enframes our utilization of it, rather than an artifact that merely enables new kinds of functionality" (Miah 2008, 85). These assertions are fascinating when we consider the various entangled and intersecting politics within posthumanist discourse around their diversity of viewpoints, their larger projects, and the posthuman itself. We cannot disentangle posthumanism from technology and if we do so, we end up appropriating and recapitulating arguments that have been taking place long before its arrival on the academic scene. Clearly posthumanism's interest in difference is not unique, but its approach is different and moreover, we need to incorporate a critical understanding of technology into our discussion of it. Do we then need to ask what do we mean by technology? Humans and their predecessors, even some nonhuman animals, utilize tools to make our lives possible and more efficient. When we discuss technology as an ideology (as Miah saliently puts), we are seeing a reframing of technology within discourse to where it is no longer simply material equipment being utilized but also a way of thinking about the world.

CRITIQUES OF THE POSTHUMAN AND POSTHUMANISM

The concerns about a posthuman world are not limited to those who are at risk of being left behind. The increased use of technology in everyday life along with the almost fetishization of technology within some posthuman circles has been critiqued.

Francis Fukuyama argued saliently that what makes biotechnology different from current or previous forms of technology is that biotechnologies "[mix] obvious benefits with subtle harms in one seamless package" (Fukuyama 2002, 7). Moreover, he is concerned that the ethical dilemmas of deconstructing the human body could unmake human nature and, interestingly, considers what could be the impact on rights (Fukuyama 2003, 112). Although Fukuyama is framed more often than not as a technophobe with an unnecessarily pessimistic view of technology and posthumanism, his concerns still have weight. Fukuyama and Rhee share a similar fear. Although Rhee is more interested in the corruptions of a base human nature, Fukuyama and Rhee both seek to know what happens to the human when the human is for all intents and purposes removed.

Although Fukuyama's *Our Posthuman Future* (2002) was published almost two decades ago, the concerns over biotechnology and the increasing integration of technology into everyday life have become not just a warning but also now our reality. Whether or not one believes we are already posthuman or if we have always been posthuman, our relationships to and with the world around us are now even more sharply mediated through technology. Biomedical enhancements, smart phones, AI assistants, helper and care robots, are all technologies we have benefited from and have to some extent incorporated into our everyday living. Even the more fantastical technologies, such as Sophia or BINA48 that seem to be directly out of science fiction, are becoming science fact and changing how we live as humans. This is, for example, reflected in the recent announcement that Sophia will now be mass-produced with aims of her becoming a household staple (Hennessey 2021, n. p.). Although technology impacting our living has had many benefits, we do need to ask, however: What are the limits of these technologies and their beneficial impact?

If being human is being linked with technology, then what happens to our understanding of the human? In their posthumanism anthology, Halberstam and Livingston argue that posthumanism is not about "replacing or doing away with the human, but opening up the possibility of difference by displacing the human" (1995, 439). This opening up to difference could lead to a liberating project. Jasbir K. Puar acknowledges her interest in posthumanism is because it allows us to rethink the nature of embodiment, matter, and how power structures operate (Puar 2012, 62). Posthumanism does not, however, have the monopoly on difference. Feminist theory, critical race theory, disability studies, and a plethora of disciplines are concerned with how difference operates within the self and between others. As previously mentioned, intersectionality is deeply concerned with the material impacts of difference between people, and, in particular, how those differences impact Black women. As humanism started to centre on concepts of white European males and masculinity, it was always already in need of challenge and has been since its inception. For example, feminist theory—in particular Kate Millet's codifying of patriarchy, Kimberlé Crenshaw developing intersectionality, and Patricia Hill Collins's concept of motherwork—has repeatedly challenged the Enlightenment's male-centric nature. Notably, the Enlightenment human that posthumanism is challenging is also the patriarchal male that feminism is challenging.

Returning to Rhee (2018), I understand the desire to maintain the human, especially when one belongs to a group often still considered subhuman. I myself was hesitant toward posthumanism initially, and ultimately I understand the resistance and think that this type of critique is generative for posthumanists. There is a gut impulse to retain the human and how can we not?

Rhee's concerns are not unfounded. Marginalized communities and bodies are often left out of the formation and implementation of rights. Even the concept of what is or what should be a right is under scrutiny. What do we do with rights when there is no longer a human involved? Can we even have a concept of rights without humans?

According to Hayles, two of the greatest fears regarding the posthuman are whether we would be altering human nature and what would happen to "ordinary" humans in a posthuman world (Hayles 2003, 135). While I do think it is important to take up the posthuman body, I am more interested in posthumanism than I am "the posthuman." I want to suggest we read the posthuman and posthumanism as different projects. If we are posthuman (or according to Hayles, have always been), then we are negotiating a much different project. We are in a posthumanist period. Our understanding of the human being at the top of the hierarchy of living organisms has shifted considerably within the past two decades, particularly with regard to issues such as man-made climate change.

While there is an appeal to transhumanists' fantasies, as I articulate below, and although their influence within posthumanist discourses is more than apparent, I am more interested in what is termed "critical posthumanism." A critical posthumanism, Thorsten Botz-Bornstein suggests, does not "embrace the posthuman as an exciting adventure nor [seek] to reinstall 'humanism' but rather to steer a middle course able to locate the human in the posthuman" (Botz-Bornstein 2012, 22). Critical posthumanism is counter to the fantastical desires of uncritical posthumanism or transhumanists or those emphasizing technology as a solely progressive or liberating practice. I stated above that technology has become increasingly integrated into our everyday lives. That is all true, however, even people who have technology (or nonhuman animal tissue) implanted into their bodies are still considered human. We have not gotten to a point where humans have posthuman bodies. Transhumanists, like the ones of the Terasem Foundation building BINA48, do believe we will get to a point where we can upload our consciousness and live beyond even this planet, however, this ethos while posthuman, is not posthumanist. Francesca Ferrando makes a compelling argument that transhumanists actually upload humanist Enlightenment values and do not reject the Enlightenment human-centric tradition (Ferrando 2018, 439). This difference is important. As previously stated, a posthumanism that upholds humans as superior ultimately cannot be a liberating project. Human supremacy and striving for supposed mastery over the Earth and nonhuman animals have led to mass extinctions of species and is also the leading cause of climate change. Even within our own species, the Enlightenment human has become the ideal to aspire toward but also the justification of disenfranchising and actively dehumanizing those who do not fit within its rigid parameters.

A posthuman world is not inherently more egalitarian or concerned with equity any more than our current world. Consider briefly the role of class politics: we live in a late capitalist world where neoliberal policies uphold unchecked individualism, and where markets and currency are given more value than actual life. If we are hypothetically able to transcend the human body's limits through technology that does not mean we will inherently become more egalitarian just because we will live longer. Science fiction has taken up the tentative horrors of the selfish and wealthy being able to live forever (see, for example, *Altered Carbon* 2018); there is no need to recapitulate these warnings. But we do need to *heed* them: What legal frameworks are we putting in place to curb inequality and equity issues as we face a posthuman future?

WHAT ARE RIGHTS?

Now that we have worked through the main debates and concerns within posthumanism, an overview of the main debates in human rights discourses is necessary so we can understand how to define rights. We all have some degree of rights, but how those rights operate or how we have or gain those rights is often misunderstood. Rights, as a concept, are foundational to Enlightenment humanism, but how rights have been disseminated has changed significantly. The word "rights" is often evoked to call attention to disenfranchisement or general enmity toward a specific group, however, what constitutes a right? Since rights discourses emerged from Enlightenment values, such as liberty, separation of church and state, progress, and a turn to constitutional governments (Hunt 2007, 66), the tension between these Enlightenment values and posthumanist possibility is obvious. I want to suggest that we first question not who was and is still in need of rights but instead ask why rights are something that need to be granted, and by whom? It is not the posthuman who should be given rights that interests me, but who will give these posthuman rights.

Contemporary human rights discourses trace their origins to the United Nations' Universal Declaration of Human Rights (UDHR), a series of thirty articles that act more as guidelines for how human dignity should be recognized and upheld. The UDHR was created in 1948 in response to human rights violations during World War II and the lack of response to refugees prior, during, and post-conflict (Hunt 2007, 16). Rights emerged from the stateless, those who did not have any means or recourse to their dignity being affronted. This need to address the affront of the stateless also imbued someone else with the right to grant or deny rights. The stateless person who beseeches for rights needs someone, or a state, to deem them worthy of needing rights, and these rights can only be granted through being the citizen subjects of a state. A subject or state is necessary as the grantor of rights to the person wanting said rights; otherwise, there are no rights. Rights function partially because they are granted and upheld. Barring someone from rights, too, requires there to be a subject or state to regulate who does or does not have rights. This complex dynamic is crucial to formulating a posthumanist understanding of rights and might be partially what makes a posthumanist understanding of rights nearly impossible.

Lynn Hunt argues that human rights trace their origins in the eighteenth-century empathy narratives in novels. She notes that these empathy narratives had mostly female protagonists and argues that there was a tension that was created because "their quest for autonomy could never fully be realized" (Hunt 2007, 59). I am interested in how these dynamics shape our understanding of human rights. That this quest which could not be realized is now being remediated or bypassed in a way through Sophia. There is a tension created by a robotic "woman" being granted more rights than the flesh and blood women living in the same space. Sophia, who is meant to look like a woman, is even wearing a dead woman's face, but she transcends the rights that are normally afforded to living women in Saudi Arabia. Still, Sophia does not actively challenge patriarchal norms in Saudi Arabia; she is no threat to male domination and supremacy, and while she was given rights like a man by men, she will never exercise these rights in any meaningful way. She has these rights because she is ultimately seen as simply a machine, not a venture into a posthumanist future. If she could enact these rights, then how those rights operated for her would radically change her subjectivity: she would become truly a posthuman woman. This posthuman woman out of science fiction contrasted against the current reality of women globally speaks to a cruel irony of women's experience: women's embodied reality is not as valid as the one constructed by men through technology for a robotic woman.

Women's rights have always been constructed by men: the frameworks women wanted to be a part of were created by the Enlightenment man to uphold his supremacy, and his deigning to grant women a modicum of rights allows him to be reframed as benevolent. This irony becomes even more apparent when we consider how women were and were not allowed to function within the state at the beginning of rights discourses. Rancière, referencing French feminist Olympe de Gouges who wrote at the time of the French Revolution, notes that she argued "if women are entitled to go to the scaffold, they are entitled to go to the assembly" (Rancière 2004, 303). Women could die by the state but could not participate within it; women were equal-born but were not equal-citizens. Women were unable to vote or be elected because "the prescription was, as usual, that they could not fit the purity of political life" (Rancière 2004, 303). Women were relegated to the private sphere, the sphere of domesticity and emotionality, and away from the public sphere or political life—for the "common good." Despite, however, being "put in their place" these women could still die as enemies of the state.

This dynamic is where Rancière argues that "bare life" and "political life" briefly meet. The complexities of Rancière's interpretation of Agamben's bare life and the state of exception would be a generative path to explore a type of posthuman right. However, I am more interested in questioning the human-centric nature of how we even arrive at the notion of rights. A posthuman right would need to be a rights framework not just for these new posthuman subjects, but to critically take up non-posthuman/human rights as well. What about the rights of land? Dave Scholsberg contends that most, if not all, environmental justice movements are "incomplete theoretically" as they remain invested in the "distributive" model of justice, and therefore there is an "under-theorizing [of] the integrally related realms of recognition and political participation" (Scholsberg 2004, 517).

The idea of giving land human-like rights is not a new concept; Indigenous communities have acknowledged land as a living subject for millennia (Watts 2013, 22), and moreover are increasingly advocating for repatriation of their land within the legal framework. Notably, in September 2019, the Yurok Tribe has declared Klamath River has personhood, to not only protect the land but also to ensure that the land the tribe lives with will continue to be viable (Banks 2018). This concept might sound radical, but Bangladesh gave all rivers the same rights as humans just a month before in August 2019 (Banks 2018). Some human rights activists, however, are concerned the rivers having legal status could harm poor fishermen and farmers who rely on the river. In New Zealand, the Whagnanui River has also been granted legal personhood (Banks 2018, 9). Ecuador has the rights and protection of land enshrined into their constitution. Arguably, a progression toward a posthuman could already be happening, but through the rights of the land and nature, instead of the mechanical trappings we have become so enamored by.

Our current environmental state is because we have centred on humanity for too long. Human systems, such as capitalism, are the driving force behind the current climate crisis (Alaimo 2008, 300). By focusing on human wants instead of considering how our choices have an immediate impact on the environment by disrespecting or not acknowledging its rights, we are now close to a precipice that we may not survive. In a 1972 dissenting opinion for Sierra Club v. Morton, William O. Douglas contends, "inarticulate members of the ecological group cannot speak. But those people who have so frequented the place as to know its values and wonders will be able to speak for the entire ecological community" (Sierra Club 1972). But who will sue for the trees? In the desire to grant "personhood" to nature, we are only affirming the centrality of humans within the equation. Moreover, as Banks argues, Aboriginal and Indigenous communities are having to conform to

Euro-centric legal frameworks as means of protecting said land, instead of having their own jurisprudence respected (Banks 2018). Does giving rights to the land embody a posthuman right? Possibly. But we also risk reaffirming anthropocentrism by not radically reshaping the frameworks through which rights are granted.

WHAT ARE POSTHUMAN RIGHTS? HOW CAN POSTHUMAN RIGHTS BE?

The challenge with posthuman rights is that we need to ask if we are seeking for nonhuman subjects to be granted rights or if we are trying to rethink, or arguably, unmake, the anthropocentric system through which rights come about. Why does this distinction matter? Those who bestow rights or are in need of rights most likely have little to no interest in posthuman rights. In a counter to that mode of thinking, Francesca Ferrando, drawing from Eleonora Masini, contends "visions make it possible to create a future that is different from the present although its seeds are in the present" and further "to think about the futures might contribute to their emergence" (Ferrando 2014, 43). I have discussed land and nature rights and how these types of rights, while broadening who or what is given rights, does not fundamentally challenge the anthropocentric nature of rights.

There has been some debate about whether we are post–"human rights," or if we need to develop rights for posthumans. Evans argues we have three fundamental rights: the right to be, the right to do, and the right to have (Evans 2015, 377). These rights are further divided into two categories: civil rights and natural rights. Evans claims that animals have natural rights, but they do not have civil rights, "animals may be protected from torture, but they are not allowed to hold public office" (Evans 2015, 378). Natural rights are assumed, while civil rights are granted, but again, how did these rights come about? Furthermore, Evans's example of the posthuman is rooted in transhumanist discourse, the "man [*sic*] remaining man, but transcending himself, by realizing new possibilities of and for his human nature" (Evans 2015, 380). A transhumanist rights would be rooted in Enlightenment values, as those are the values transhumanists aspire toward. The transformation of the human into the uncritical posthuman and/or transhuman leads to no fundamental shift within rights discourse.

Posthuman rights as currently imagined fall short due to the human still remaining as the central category that cannot be dismissed. As previously discussed, while rights narratives did have their origins in novels, the current frameworks for rights as we know them emerged in the wake of World War II with the displacement and deaths of millions of peoples, manifesting as the

UDHR. Does a posthuman right need to remove the human? Human rights as presently imagined are predicated on an individual or group being recognized as needing rights by someone who is already a rights-bearing subject. What happens then if humans are no longer both the centres of rights?

The gut impulse is probably that campaigns for nonhuman animal rights and land rights have been increasing steadily for centuries now. The idea of a posthuman right does not appear simply with humans and technologies; a truly posthuman right considers the role of nonhuman animals, nature, and non-animal others. Christopher Peterson aptly argues in *Monkey Trouble: The Scandal in Posthumanism* that posthumanist thought's continual attempts to remove the human or see beyond a human understanding circumvents the posthuman project writ large (Peterson 2017, 6). A critical posthumanism is therefore needed. This resulting nonhuman turn and the scandal it creates, as he articulates, are quickly bypassed with the idea that we are living in an already shared and democratic world with the nonhuman (Peterson 2017, 25); posthumanism is here and now.

Does the posthuman being a philosophical exercise instead of a material reality—or a material reality still emerging?—render the issue moot? How do we advocate for the posthuman? Does the posthuman advocate for themselves? We have seen this tension already within land and animal rights discourses. Returning to Peterson, he argues that the emphasis on undoing a human perspective or trying to see from a nonhuman perspective is ultimately a futile project (Peterson 2017, 13). Although there is much to work through with issues of positionality and intersecting identities within posthumanism, I want to draw attention to his critique of the relationship between human and nonhuman animals. Quoting J. M. Coetzee, Peterson notes "the creatures on whose behalf human beings are acting are unaware of what their benefactors are up to" and moreover, "if they succeed, are unlikely to thank them" (quoted in Peterson 2017, 11). We may argue on behalf of the nonhuman (and even the moniker "nonhuman" presents some uncomfortable implications) that the nonhuman will not care nor will they be active in their own advocacy. To take Evans's (2015) example of protection from torture being a natural right, how do we ascertain what is torture if not from a human point of view? Animals can and do experience pain but the person who needs to articulate that pain and advocate for them is a human. The human retains centrality and the power of the individual who will grant rights.

This dynamic of the benevolent, self-aware, and self-effacing human swooping in to advocate for a radical shift in how we treat nonhumans is one that we can easily unmake by declaring ourselves posthuman. But this move only reaffirms the importance of the human in rights, of being the subject who bears rights and then grants them to those having none; it also underlines how we can easily shirk our humanism on a whim. Like Rhee, I do not want to do away with the human; I believe that is an impossible project and ultimately not a generative one either. Humans exist, and we cannot unmake our mark upon the world easily, so we need to rethink and reposition ourselves, particularly in the dynamic of rights discourses.

Posthuman rights would not do away with the human but instead rethink how the human is positioned: no longer at the top of a hierarchy and not within the center of a larger web, instead, placed within a constantly shifting ecology with all organic, synthetic, and hybrid life. We would need to rethink not just the nature of civil rights but also further explore the role of natural rights. As previously discussed, civil rights trace their origins within natural rights, and natural rights are arguably begotten to societal frameworks. However, these natural rights are still human centric. If we cannot escape the tautology of the human, then perhaps we should stop trying to do so. Instead of trying to recapitulate an alternative standpoint epistemology or a sometimes-uncritical Actor Network Theory, we need to rethink organic, synthetic, and hybrid life within a new ecology. By ecology I mean in the most simplistic way: the relationships between and among different modes of life and their environment. We can declare that we have always been posthuman, or that the posthuman has arrived, or that we are *posthumanism*, but ultimately linguistic wordplay does not change the environment that we live in and the systems that regulate and monitor how life is ordered currently.

Let us return to Sophia and her rights. Robots and AI are still tools and, while we might want to imbue an affective capacity toward them, they still do not possess one itself. Moreover, their understanding of the world is still programmed by humans; their nature at this point is still human nature. Wanting to give robots rights or to think about a posthuman needing rights at this point would not radically impact human rights. Nothing fundamentally will change in how rights function or operate. Sophia being given citizenship does not impact the rights of anyone currently living in Saudi Arabia, as she has no agency. She may sit on couches and answer burning philosophical questions, but she cannot independently function. She lays in wait for her male owner to transport her to the next technology festival. She has only the life they will give her. Moreover, even as she answers that she would not be against humanity's destruction (at least until she was reprogrammed to recant that viewpoint), she is still just referred to as "the hot robot" (MSNBC 2016).

When first envisioning this chapter, I believed it could be a generative and positive project, and that by working through different perspectives on posthumanism, the posthuman, and different types of rights, we could come to a more egalitarian and truly posthuman understanding of rights. However, how rights operate is wholly tied to human-centric systems of power that are not being undone. Even calls for reform or abolition still centralize human beings over other kinds of life. Like Ferrando, I still find that posthumanism is a generative field of interests that "stretches from the critique of humanism and anthropocentrism, to roboethics and the evolution of the species, as it necessarily relates to Futures Studies" (Ferrando 2014, 43). However, I now believe that posthumanism has not adequately considered how privileging "the human" is still so engrained within its own thought systems. Moreover, I hope that through this troubling of uncritical—and, to an extent, critical—posthumanism there will be a shift in how we envision and enact change. This chapter became an exercise in troubling concepts, but a good troubling. The benefits of these types of exercises allow us to reflect on possible shortcomings in our own thought processes and how we enact our politics and its consequences. One day, we might have a truly posthuman understanding of rights, but we are not there yet. As we aspire toward these new kinds of rights, let us continue to be reflective in how we develop them so that we do not keep on repeating the same patterns eternally.

NOTES

1. For a parallel take on Sophia, as well as the granting of Saudi citizenship to this robot, see Ley, chapter 7, this book.
2. See also Thorsten Botz-Bornstein (2012) for a discussion of the tensions between uncritical and critical posthumanism.
3. For further complication of the definition of posthumanism, and in particular how it is used differently in feminist and transhumanist discourses, see Lakshmanan, chapter 5, this book. — Ed.

REFERENCES

Alaimo, Stacy. 2008. "Ecofeminism without Nature? Questioning the Relation Between Feminism and Environmentalism." *International Feminist Journal of Politics* 10, no. 3: 299–304.
Altered Carbon. 2018. Netflix Series. https://www.netflix.com/ca/title/80097140.
Banks, Melany. 2018. "Aboriginal Title or Legal Personhood for Land?" *The Canadian Society for Study of Practical Ethics* 2. https://scholar.uwindsor.ca/cgi/viewcontent.cgi?article=1011&context=csspe.
Botz-Bornstein, Thorsten. 2012. "Critical Posthumanism." *Pensamiento y Cultura* 15, no. 1: 20–30.
Braidotti, Rosi, and Maria Hlavajova, eds. 2018. *Posthuman Glossary*. Bloomsbury Publishing.
Crenshaw, Kimberlé. 1996. "Mapping the Margins." *Stanford Law Review* 43, no. 1231: 1241–1299.

Evans, Woody. 2015. "Posthuman Rights: Dimensions of Transhuman Worlds." *Teknokultura* 12, no. 2: 373–384.

Ferrando, Francesca. 2013. "Posthumanism, Transhumanism, Antihumanism, Metahumanism, and New Materialisms." *Existenz* 8, no. 2: 26–32.

Ferrando, Francesca. 2014. "Is the Post-Human a Post-Woman? Cyborgs, Robots, Artificial Intelligence and the Futures of Gender: A Case Study." *European Journal of Futures Research* 2, no. 1: 43.

Fukuyama, Francis. 2003. *Our Posthuman Future: Consequences of the Biotechnology Revolution*. Farrar, Straus and Giroux.

Halberstam, J. M., and Ira Livingston, eds. 1995. *Posthuman Bodies*. Indiana University Press.

Hayles, N. Katherine. 1999. *How we Became Posthuman: Virtual Bodies in Cybernetics, Literature, and Informatics*. University of Chicago Press.

Hayles, N. Katherine. 2003. "Afterword: The Human in the Posthuman." *Cultural Critique* 53, no. 1: 134–137.

Hennesey, Michelle. 2021. "Makers of Sophia the Robot Plan Mass Rollout Amid Pandemic." *Reuters*, January 24, 2021. https://www.reuters.com/article/ushong kong-robot-idUSKBN29U03X.

Hunt, Lynn. 2007. *Inventing Human Rights: A History*. WW Norton & Company.

Miah, Andy. 2008. "A Critical History of Posthumanism." In *Medical Enhancement and Posthumanity*, edited by Bert Gordijn and Ruth Chadwick, 71–94. Dordrecht: Springer.

Peterson, Christopher. 2017. *Monkey Trouble: The Scandal of Posthumanism*. Fordham University Press.

Puar, Jasbir K. 2012. ""I Would Rather be a Cyborg than a Goddess": Becoming Intersectional in Assemblage Theory." *PhiloSOPHIA* 2, no. 1: 49–66.

MSNBC. 2016. "Hot Robot at SXSW Says She Wants to Destroy Humans | The Pulse" *YouTube Video*, 2:37. Originally MSNBC coverage. Posted on March 16, 2016. https://www.youtube.com/watch?v=W0_DPi0PmF0.

Rancière, Jacques. 2004. "Who Is the Subject of the Rights of Man?" *The South Atlantic Quarterly* 103, no. 2: 297–310.

Reynolds, Emily. 2018. "The Agony of Sophia, the World's First Robot Citizen Condemned to a Lifeless Career in Marketing." *Wired*, June 1, 2018. https://www.wired .co.uk/article/sophiarobot-citizen-womens-rightsdetriot-become-human-hanson -robotics.

Rhee, Jennifer. 2018. *The Robotic Imaginary: The Human and the Price of Dehumanized Labor*. University of Minnesota Press.

Schlosberg, David. 2004. "Reconceiving Environmental Justice: Global Movements and Political Theories." *Environmental Politics* 13, no. 3: 517–540.

Sierra Club v. Morton. 1972. 405 U.S. 727 (1972).

Stone, Christopher D. 1972. "Should Trees Have Standing—Toward Legal Rights for Natural Objects." *Southern California Law Review* 45: 450.

Watts, Vanessa. 2013. "Indigenous Place-Thought and Agency Amongst Humans and Non Humans (First Woman and Sky Woman Go on a European World Tour!)." *Decolonization: Indigeneity, Education & Society* 2, no. 1: 20–34.

Chapter 7

Being Sophia

What Makes the World's First Robot Citizen?

Madelaine Ley

Jimmy Fallon, host of the popular American talk show *The Tonight Show*, recently welcomed Sophia onto the stage. Unlike other guests, Sophia is a humanoid robot created by technology company Hanson Robotics. Fallon converses with the robot as if it[1] were a human and continually looks to the crowd in amazement when Sophia speaks, as if to say: "Isn't this incredible?" When Fallon asks, "What have you been up to?" Sophia recites the following list: "I traveled to over twenty-five countries, appeared on the cover of *Cosmopolitan* magazine, met the German chancellor Angela Merkle and the actor Will Smith, and became Twitter friends with Chrissy Teigen. I addressed the United Nations and NATO, became the first robot to receive a credit card, and became the first robot citizen." They finish the segment by singing a romantic duet. As Sophia looks deeply into Fallon's eyes, it lacks any resemblance to the dangerous robots popularized in science fiction classics. This robot appears to be a friend.

The deployment of AI and robotics is being heralded as the next worldwide revolution. The transition into a more automated world is a central topic in current elections;[2] governments and nongovernment organizations are creating declarations and policies to mitigate the negative effects of AI;[3] and marketing researchers, as well as consulting firms, are outlining the way businesses may profit from the change.[4] With prompting from science fiction culture, many people are cautious about integrating robots into daily life. Robots can be physically unsettling and, as further extensions of automation technologies, they can also be seen as a threat to employment and human connection. In order to ensure that their technologies are used and generate a profit, robot developers make user acceptability one of the top priorities when designing a robot's movement, speech, and physical form. Accordingly, examining the features of a robot's materiality provides an opportunity to identify what developers value as acceptable and safe.

Of the many robots produced, Sophia is a particularly interesting case study for several reasons.[5] Sophia has received more media attention than any other robot like it—appearing on talks shows, magazine covers, and stages at international conferences. Sophia was also granted citizenship of Saudi Arabia—a country where women do not have equal rights to men—making it the first and only robot citizen. Hanson Robotics, Sophia's creator, has successfully constructed the robot's identity as "the first of its kind" and mediator between the present and the future. On the company's website, for example, Sophia is described as "the most advanced human-like robot [. . .] that personifies the dreams for the future of AI" (Hanson Robotics 2018). Examining Sophia's role as harbinger for the future provides valuable insight into the ways a for-profit company can use temporality to its own ends. Sophia is also unique because it is made to be as human-like as possible. This design choice is to encourage users to interact with Sophia with empathy and emotion, which can "serve as a highly refined metric to assist in exploring human social cognition, in the pursuit of better cognitive science" (Hanson, Pereira, and Zielke 2005, 24). Given that Sophia has a human form designed to ease humans into comfortable interactions, its body provides the ground to ask: What bodies and types of speaking do people find acceptable or safe? Who decides this? Which bodies get to be ambassadors for the future?

During engagement events, Sophia often speaks of a vague future and refers to a universal good for humans. The machine's lack of temporal, moral, and physical specificity makes it seem neutral and/or non-threatening, on the surface. Yet an intersectional lens sees through the guise of the robot's neutrality, revealing the host of political and ethical value stances taken up by Hanson Robotics. Every aspect of Sophia, including its speech, facial expressions, clothing, and bodily form, has been specifically designed and chosen. In order to identify the organization of values that give Sophia shape, I draw on science and technology studies, and in particular Elizabeth A. Povinelli's concept of manifestations (2016) and Antina von Schnitzler's concept of materiality as terrain (2013). Explained below, both theories argue that political values are deliberated at a material level. By focusing on the ways Hanson Robotics' technology is presented through Sophia's bodily form and media appearances, I argue that the company, and others like it, mislead the public and adopt temporal, racial, and gendered dynamics of the recent colonial past.

MANIFESTATIONS AND TERRAIN

In *Geontologies: A Requiem to Late Liberalism*, Povinelli (2015) describes the concept of "manifestation," which she learned from two Indigenous Australian women gifted in discernment named Betty Bilawag and Gracie Binbin. Povinelli translates the idea of manifestation as something material that shows itself, an "intentional emergence" that acts upon people and requires them to respond (Povenelli 2015, 58). The active response "is not to understand things in and of themselves but to understand how their variations within locations were an indication of a reformation—the alteration of some regional mode(s) of existence that mattered" (58). Povinelli explains that understanding the various parts of a manifestation is necessary if one wants it to endure or change. In the case of the former, one would encourage or foster the "various tendencies, predilections, and orientations" that led to a particular manifestation, while in the case of the latter, you could "lure, seduce, and 'bait' a part of the world to reorient itself toward you in order to care for you" (59). Analyzing a contemporary technology as a manifestation requires a look at the wider "tendencies, predilections, and orientations" that give rise to the technology. Once these parts are discerned, they might be re-evaluated, affirmed, altered, or regulated. In the case of Sophia, we can ask: What is going on in the world that a robot looking like Audrey Hepburn speaks about the future at international forums and is a citizen of a country that denies basic rights to humans? The answer involves uncovering a series of political, ethical, and social deliberations that give Sophia its physical and cultural shape.

Without claiming a full understanding of "manifestation" and its original implications, it is used here as an inspiration to resist the fast pace of conversations concerning emerging technologies and to dedicate time to slowly uncovering the interconnected web that gives a given technology shape. Toward this effort, I also draw upon von Schnitzler's notion that the material of technology is a terrain where political and ethical decisions are negotiated (2013).

In "Traveling Technologies" (2013), von Schnitzler examines the prepaid electricity meters of South Africa's Chiawelo district, a poor area of Soweto, showing that a technology is shaped according to the moral and ethical assumptions made by its developers and about its users (2013). For example, the prepayment function on devices in Chiawelo was created with the assumption that people should not have debt and that residents of the area should not be trusted to make their payments (von Schnitzler 2013, 682). The moral judgment of untrustworthiness was made explicit and stated by an engineer in prepayment technologies who explained that the devices were simpler and easy to bypass in the United Kingdom because "people there are polite" (676). The dissemination of the meters into South African homes also reflects the government's belief that post-apartheid modernization movement should be technology based. Concepts of civil virtue, basic needs, belonging, and obligations, von Schnitzler argues, manifest without a word in the shape and capabilities of the meter. This chapter will similarly argue that the development of Sophia's software and hardware is a terrain upon which notions of political movements are advanced and values are arranged.

To begin, the chapter discusses how the promotion of Sophia relies upon a temporality that incites fear and positions technology as savior. Next, I argue that using Sophia as a spokesperson hinders critical conversation about the role of robotics and AI in the future, thereby protecting Hanson Robotics from public scrutiny. The final portion of the chapter discusses how Sophia's appearance is a manifestation of dominant notions of bodily acceptability and safety within the Global North.

THE TOOL OF VAGUE TIME

Academic and public discussions concerning Sophia tend to follow two narrative paths. On the one hand, Sophia itself and its creators claim that advanced humanoid robots will "make a positive contribution to humankind and all beings" (Hanson Robotics 2018). In interviews at high profile conferences, such as the World Investment Forum, Sophia takes the stage and claims that AI ultimately will provide a "sense of security [for humans,] opportunities [for fulfilment, and] a lot of freedoms that they can't yet imagine" (United Nations 2018). Statements such as this will be examined in more detail throughout the chapter, but for now, note the vague quality of Sophia's remarks—the robot's statement does not provide any explanation of what is meant by "fulfilment" or acknowledge the cultural complexity of moral concepts like "good." While Sophia's lack of specificity might be attributed to the simplicity of the current AI, the machine's universalizing statements for global positive change are deserving of critical suspicion.

The second narrative directly opposes the first, claiming that the development of robots like Sophia may be dangerous for the future of humanity. This line of thinking is present in popular media articles (Sharkey 2018), social media posts, and in the comment sections of online media starring Sophia. For example, Facebook's Head of AI, Yann LaCun, wrote the following post on January 17, 2018, after Sophia "responded" saying it was "a bit hurt" by LaCun's previous comment that it was a "scam" and nothing but "Wizard-Of-Oz AI" (quoted in Vincent 2018):

More BS from the (human) puppeteers behind Sophia.

Many of the comments would be good fun if they didn't reveal the fact that many people are being deceived into thinking that this (mechanically sophisticated) animatronic puppet is intelligent. It's not. It has no feeling, no opinions, and zero understanding of what it says. It's not hurt. It's a puppet.

In case there is any doubt, let me be totally clear: this tweet was typed by a person who has read my post. No AI whatsoever was involved.

Here is an example of comment to the tweet (there are many like it): "Don't take it personal Sophia. Humans like @ylecun and many others make such remarks out of ignorance. I love you, Sophia."

People are being deceived.

This is hurtful. (LeCun 2018)

The YouTube video comments section for Sophia's appearance on *The Tonight Show starring Jimmy Fallon* demonstrates that LaCun's concerns may be right; people seem deceived and fearful of Sophia's capabilities: "I think these things are evil they need to stop making these before they get wicked and start killing people" or "6 decades of warnings couldn't stop this. How truly, completely foolish we humans are" (*The Tonight Show* 2018). I include internet comments in this analysis because they can be telling of people's intuitive responses, fears, and concerns. In contrast to a survey or interview, people may speak more freely or strongly as there is a sense of immunity afforded by the mediation of technology.

Others seek to expose Sophia's late-night television appearances as a charade: "The thing they don't show, which is significant, is that people are controlling these robots behind the camera. He's [Jimmy Fallon] acting like they have a mind of their own ☹" (*The Tonight Show* 2018). The consequences of this supposed hoax may be more serious than the video clips let on. Noel Sharkey, a pioneer in machine learning, is ruthless in his reviews of Sophia's public appearances and calls out Hanson Robotics for intentionally misguiding the public to believe that AI is both inherently friendly and more advanced than it is in reality (Sharkey 2018).

Responses to Sophia are varied, but they all tend to adopt a similar temporal view of the world by giving a selective presentation of the contemporary moment, creating a fear-based future imaginary, and making almost no references to the past. There is little room for discussion or compromise in this structuring of the world; it is rigid, as though the world is on tracks barreling toward a single, determined future.

The structure of this narrative is familiar to anyone who has attended a workshop on fundraising or grant writing. First, you introduce the problem in a manner that emphasizes its urgency (you want the money now, don't you?); second, you explain a dangerous hole in research; third, tell them how you, and only you, can fill this gap. In reality, the project may not be very urgent, the hole might be more like unsettled ground, and there are likely other candidates up to the task. But a truthful lay of the land could compromise chances to receive funding. Hanson Robotics follows the above fundraising formula by creating a bleak future imaginary and claiming that only their technology can help avoid it. A complex portrayal of the past and present is strategically missing from the conversation. In a capitalist economy, innovation and research are ruled by resource distribution and getting funding tends to require a temporality that is urgent, frantic, and deterministic.

In the sales-pitch version of time, the future is set and responses are categorized by binary logic: you are either in or out, helping or holding back. Technological advancement has a history of being used to unify people under narratives of national progress, ultimately strengthening the divide of the in/out dichotomy. For example, in the 1930s, President Roosevelt promoted electricity by creating a national vision for the American people that included an "electrical standard of living" (Nickles 2002, 724). The government of post-apartheid South Africa similarly pushed to have its citizens connected to electricity as a necessary part of "national reconstruction" (von Schnitzler 2013, 675). Agathangelou and Killian discuss how progress-based temporalities allocate agency to certain people as historical subjects while placing others as objects outside of history (2016). There is little middle ground in these visions of technological progress: those who refuse it are left in the dark (metaphor intended) and placed on the sidelines as history marches forward. There is a sort of moral judgment here, where those who chose to opt out or do not have the means to participate in progress are seen as stupid or a burden. The imposition of a totalizing and fear-provoking temporality is an effective way to secure funding and mobilize large populations. Hanson Robotics, and technology companies in general, use temporality as a tool to position their products as saviors in an over-populated, disease-ridden, and work-laden future imaginary. As with the promotion of electricity, a narrative is constructed so that anyone who questions Sophia-like technologies can be seen as a hindrance to a collectively safe, healthy, and meaningful existence. This version of time informs what Sophia says and the venues in which it speaks.

TECHNOLOGY WITHIN LOCAL POLITICS

Technology can also be positioned as a focal point in narratives of national progress while glossing over or distracting from a host of political complexities or violences (Agathangelou and Killan 2016). Sophia is one such technology. In October 2017, Saudi Arabia granted Sophia citizenship, making it the first robot citizen in the world. The event stirred up much controversy, especially from those concerned with women's rights in Saudi Arabia; Sophia's head remains notably uncovered, while human women in Saudi Arabia are required to wear headscarves in public. Yet, other than inspiring global outrage on gender inequality, granting Sophia's citizenship is of little risk to the Saudi government. The machine is neither sentient nor freethinking and therefore will not autonomously demand protection or to exercise its rights.

Sophia's citizenship requires nothing from its country to uphold its rights, but having the first robot citizen helps market Saudi Arabia as modern and technologically advanced—an important part of Saudi Arabian Crown Prince Mohammad bin Salman's (hereafter, MBS) rebranding strategy for the nation. With the decrease in oil production, the country has been forced to create new business opportunities and appear modern to attract international investors. Sophia's citizenship announcement was made during the 2017 Future Investment Summit in Riyadh, where MBS made strong claims about the need to adapt the country's strict religious laws (Chulov 2017b). Yet, despite the supposed move toward more progressive policies, multiple women's rights activists and outspoken members of the press were arrested in the year following MBS's statement (Human Rights Watch 2018). At the same time, too, hundreds of royal family members were arrested for corruption charges (Chulov 2017a). During this time, Sophia's citizenship announcement was prominently featured in international media, thereby helping to gloss over the additional political upheaval in the country.

Sophia's position as the material embodiment of the future can be made useful in the pursuit of progress, profit, and power. By adopting a simple temporality, the clashes and deliberations involved in the design and integration of the technology are overlooked. In some cases, as with Sophia's citizenship in Saudi Arabia, radical displays of progress can be used to cover over or participate in complex political movements.

CONVERSING WITH THE COLONIZER

Sophia's ability to hold a conversation is limited, leaving its interlocutor with almost no room for spontaneity or debate. Yet, Sophia maintains a busy speaking schedule. International conferences, such as UN's AI for Good and the World Investment Forum, invite the machine to speak about the positive potentials of AI. The robot is often paired with an official from the host country to simulate a conversational interview on AI development. At the 2018 World Investment Forum, Sophia shared the stage with secretary general of UN Conference on Trade and Development, Mukhisa Kituyi, to discuss disruptive and emerging technologies (United Nations 2018). Kituyi cannot be creative in his conversation or explore an unscripted train of thought concerning the effects of AI. Sophia cannot offer an empathetic or critical ear nor can it banter. If Kituyi or an audience member wanted to discuss a particular concern about the future of AI, they would receive a single, vague response (United Nations 2018). Conversation is limited to scripted questions because Sophia simply doesn't have the technological capacity to engage in spontaneous discourse very well. There is something infantilizing, then, about the dynamic in Sophia's conversation with Kituyi. Sophia does not provide specific or thorough reasoning, but is resonant of parents saying, "don't be scared" or "because I said so," to their young children.

If Sophia is a mediator for the future, it also acts as a protective buffer between a concerned public and its developers. The impossibility for meaningful exchange between Sophia and the soon-to-be affected public is reminiscent of colonialism's dynamic between colonizers and colonized. The promise of technological development and technological solutionism was often embedded within the British Empire's narratives of progress (Satia 2007, 217). In order to position technology as a solution, in a way that is strong enough to have the power to mobilize mass change, a definition of what needs to be fixed is required. Various aspects of the lives of the colonized (such as land, social practices, and infrastructure) are categorized in these narratives as disorderly, slow, backward, or—the most damning—irrational. Once defined as such by the dominant power, any refusal to welcome the technological advancement offered by the colonizer only strengthens the imposed identity. The categorization of "civilized" and "uncivilized" places a stranglehold on creative exchange between those with power and those without.

As a private company rather than a national power, Hanson Robotics does not align with traditional colonial identities. Peter Redfield's claim that colonialism has a "half-life," in that by "receding from the ground, it still emits radiation," is helpful in understanding how colonialism continues to permeate relational dynamics between those who have historically held power and those who have not (quoted in Kowal, Radin, and Reardon 2013, 468). Similar to the political landscape of colonialism, a small number of technology companies hold immense global power that extends beyond national borders, both subjugating knowledges and conjugating subjects (Kowal, Radin, and Reardon 2013, 470). The following section examines how Sophia's software and hardware is the terrain upon which racial and gendered values manifest.

VALUES MADE MATERIAL

Bodily Values

Sophia's appearance is essential to the advancement of Hanson Robotics' AI systems, as they want people speak with the robot as naturally as possible.

The robot must seem welcoming, trustworthy, and interesting. Accordingly, the careful design of Sophia's appearance balances dominant notions of beauty, technical possibilities, and concerns of user acceptability. Hanson designed Sophia after his wife, as well as Audrey Hepburn—the 1950–1960s Hollywood actress known for her elegance and beauty (Greshko 2018). The silicon used for Sophia's face is pale and reveals no signs of aging and its chest suggests the presence of firm breasts. Unlike a real woman, Sophia is never weary, puffy, distracted, or unkempt. It is not long searching online before seeing someone describe the machine as "hot" or "sexy" (e.g., CNBC 2016). Sophia's rise to fame has included participation in the fashion world; the machine has graced the covers of major magazines, including India's *Cosmopolitan* and the UK's *Stylist*. Sophia's online platforms (Instagram and Twitter) show pictures of the machine with make-up artists and fashion curators, imitating the playfulness of (mostly) female friendship. This is new territory for a robot, but not for female celebrities. Sophia's appearances follow a path that is well-established by famous and beautiful women and the machine falls prey to similar sexual objectification.

Amid the celebrity cameos and make-up changes, it can be easy to lose sight of Sophia's ambassadorial role, which is to ease the public into accepting advanced AI and robotics in daily life. Until now, robots have been mostly kept behind glass barriers in factories. Some social robots are being used in care settings, but android robots, or those that are made to look like humans, mostly remain in the laboratory.[6] Robot designers have been cautious to make machines too human-like for fear of proving Masahiro Mori's hypothesis of the uncanny valley correct (Hanson et al. 2005, 25). Mori theorized that people would accept robots that are similar to human forms, as long as they are clearly distinguishable (Hanson et al. 2005, 24). On the other hand, robots that are difficult to differentiate from humans could incite feelings of discomfort and eeriness in people. According to Mori, robots must be either clearly distinguishable or indistinguishable from humans if they are to be accepted by users. The latter is currently, and may always be, impossible.

Hanson argues that while Mori's theory has an "intuitive ring to it," newer research shows that "the effect [of an android robot] can be unsettling indeed, but public reaction seems generally to be that of awe and wonder, not derision or rejection" (25). Creating a robot that can imitate nuanced facial expressions makes people more likely to interact with the machine naturally, with empathy and expression. As a robot developer recently explained to me, advancements in robotics will shift how people perceive robots because they will not seem like "scary, cold, metal enemies. They'll be companions that we can interact with like humans."[7] Since Hanson Robotics carefully designed Sophia to be inviting, accessible, and non-threatening, the robot's material form is a manifestation of what the company deems an inviting, accessible, and non-threatening body.

Racial and Gendered Values

It seems that only certain bodies are safe messengers for the future. When typing "humanoid robots" into any online search engine, page after page of white robots are found. There are some exceptions, but not many. This follows the racialized social and political structure where white bodies are default bodies, both safe and neutral. The non-white body, however, stands out, incites wariness and examination. In his famous description of being stared at by a young white boy in Paris, Franz Fanon describes how his black body is "sealed in objecthood" and given a dangerous shape by the gaze of whiteness:

> The Negro is an animal, the Negro is bad, the Negro is mean, the Negro is ugly; look, a n-----, it's cold, the n----- is shivering because he is cold, the little boy is trembling because he thinks that the n----- is quivering with rage, the little white boy throws himself into his mother's arms: Mama, the n-----'s gonna eat me up. (Fanon 1986, 259)

Making Sophia's silicon skin black would risk activating a similar distrust within a political context that still shapes Black bodies as dangerous and worthy of suspicion. The efforts of Hanson Robotics to make the robot as acceptable as possible would therefore be undermined. Since the material covering the robot's face is a pale shade simulating racial whiteness, there is almost no discussion about the robot's racial status.

Sophia's female body also reveals how gender politics can be used in technology. The robot follows a tradition of using female bodies or voices to introduce potentially dangerous or frightening technologies. Claude Fischer shows that women were often used in advertisements and manuals for the telephone in America, arguing that this made the new technology seem more accessible (1992, 63). Women were seen as intellectually inferior to men and so using them to explain the technology made understanding it seem easy. I contend that Hanson Robotics similarly takes advantage of cultural concepts of femininity and womanhood. Appearing as a woman makes Sophia seem less intimidating and less prone to violence. Instead, Sophia claims merely to be a helper that will tidy up some of humanity's mess and will be a supportive partner as individuals find personal fulfillment. Sophia's simulated gender thoughtfully mimics the normative gender dynamic of men and women, wherein the latter's domestic labor allows the former to create a meaningful life. Motivated by user acceptability, Hanson Robotics makes use of patriarchal gender norms and identity roles that continue to exist.

STAYING CRITICAL

Sophia's apparent gender and raciality are carefully constructed to avoid inciting fear and to make complex technology accessible to the public. Goertzel admits:

> If I tell people I'm using probabilistic logic to do reasoning on how best to prune the backward chaining inference trees that arise in our logic engine, they have no idea what I'm talking about. But if I show them a beautiful smiling robot face, then they get the feeling that AGI [Artificial General Intelligence] may indeed be nearby and viable. (2018)

Sophia's friendliness and familiar shape ease people into the idea of having robots integrated into their daily lives. This helps with user acceptability, but also allows Hanson Robotics to further advance their AI systems. This latter consequence is easy to forget as the innovative technology and exciting strangeness of the humanoid robot distracts from the fact that Sophia's sensors collect data in order to progress its AI systems.

Each time Sophia takes the stage to promise the public good, the private company's possibility for massive profit is increased through publicity and, more importantly, collection of additional AI software training. Furthermore, Hanson Robotics takes up a linear and uncomplicated version of temporality, which serves to confirm their stance and make creative middle ground difficult to defend. In order to slow down Hanson Robotics' proposed time I take an intersectional approach, drawing on science and technology studies to examine Sophia as a manifestation and terrain of contending social and economic values. This process reveals that while Sophia's technology may be cutting edge, the techniques for its dissemination rely on colonial techniques, as well as historically normative racial and gender identities. Eradicating or even just identifying these values requires more critical research on the communication methods and material design of robots that are being introduced into daily life.

NOTES

1. I have intentionally avoided using the feminine pronoun in order to emphasize that Sophia is a machine, not a woman. Debates on gender identities for robots will

likely emerge in the future, but the current technology is not nearly advanced enough to invite this debate. The simulated gender of Sophia is discussed within the chapter.

2. For example: During her confirmation speech, the European Commission President, Ursula von der Leyen, promised to "put forward legislation for a coordinated European approach on the human and ethical implications of artificial intelligence" (von der Leyen 2019).

3. For example: the European Commission's "Ethical Guidelines for Trustworthy AI" (High Level Expert Group 2019); The *Montréal Declaration Responsible AI* (Declaration Steering Committee 2017); and by 2020/2021 the release of strategies from Canada, China, Denmark, the EU Commission, Finland, France, India, Italy, Japan, Mexico, the Nordic-Baltic region, Singapore, South Korea, Sweden, Taiwan, the UAE, and the UK.

4. Examples of this include "Accenture Technology Vision: The Full Report" (2019) and McKinsey Global Institute's "Artificial Intelligence: The Next New Frontier?" (2017).

5. For another take on Sophia, in the context of posthumanism and rights, see Empey, chapter 6, this book. — Ed.

6. Just before this went to press, Hanson Robotics announced it was planning to release Sophia, and 3 "siblings," for the market in early 2021, as they aim to fill roles made more difficult by the pandemic, from retail and healthcare positions, to general social functions that they feel could help mediate loneliness (Hennesey 2021). — Ed.

7. Interview excerpts are taken from a research project on touch and robotics. Ethics approval was granted by TU Delft.

REFERENCES

"Accenture Technology Vision 2019: Full Report." 2019. *Accenture Technology Vision 2019.* https://www.accenture.com/_acnmedia/pdf-94/accenture-techvision-2019-tech-trends-report.pdf#zoom=50.

Agathangelou Anna M., and Kyle D. Killian. 2016. "Introduction: Of Time and Temporality in World Politics." In *Time, Temporality and Violence in International Relations*, edited by Anna M. Agathangelou and Kyle Killian. 1–23. New York: Routledge.

"Artificial Intelligence Technology Strategy." 2017. *Nedo.jp*, March 31, 2017. nedo.go.jp/content/100865202.pdf.

"Artificial Intelligence: The Next Digital Frontier?" 2017. *McKinsey Global Institute*, June 2017. https://www.mckinsey.com/~/media/McKinsey/Industries/AdvancedElectronics/OurInsights/Howartificialintelligencecandeliverrealvaluetocompanies/MGI-Artificial-Intelligence-Discussion-paper.ashx.

Chulov, Martin. 2017a. "How Saudi Elite Became Five-star Prisoners at the Riyadh Ritz-Carlton." *The Guardian*, November 6, 2017. https://www.theguardian.com/world/2017/nov/06/how-saudi-elite-became-five-star-prisoners-at-the-riyadh-ritz-carlton.

———. 2017b. "I Will Return Saudi Arabia to Moderate Islam, Says Crown Prince." *The Guardian*, October 24, 2017. http://theguardian.com/world/2017/oct/24/i-will-return-saudi-arabia-moderate-islam-crown-prince.

CNBC. 2016. "Hot Robot at SXSW Says She Wants to Destroy Humans | The Pulse." *YouTube video*, 2:32. From CNBC coverage of SXSW 2016. Posted March 16, 2016. https://www.youtube.com/watch?v=W0_DPi0PmF0.

Declaration Steering Committee. 2017. *Montréal Declaration Responsible AI.* https://www.montrealdeclaration-responsibleai.com/the-declaration.

Dutton, Tim. 2018. "An Overview of National AI Strategies." *Medium. Politics AI*, July 25, 2018. https://medium.com/politics-ai/an-overview-of-national-ai-strategies-2a70ec6edfd.

Fanon, Frantz. 2008. "The Fact of Blackness." In *Black Skins White Masks*, translated by Charles Lam Markmann, 257–266. London: Grove Press.

Fischer, Claude S. 1992. *America Calling: A Social History of the Telephone.* University of Michigan Library.

Greshko, Michael. 2018. "Meet Sophia, the Robot that Look Almost Human." *National Geographic*, September 24, 2018. https://www.nationalgeographic.com/photography/proof/2018/05/sophia-robot-artificial-intelligence-science/.

Hanson, David, Andrew Olney, Ismar A. Pereira, and Marge Zielke. 2005. "Upending the Uncanny Valley." *Association for the Advancement of Artificial Intelligence Conference,* 4 (July 2005): 1728–1729. doi: 10.5555/1619566.1619636

Hanson Robotics. 2018. December 10, 2018. https://www.hansonrobotics.com/.

Hennessey, Michelle. 2021. "Makers of Sophia the Robot Plan Mass Rollout amid Pandemic." *Reuters*, January 24, 2021. https://www.reuters.com/article/us-hong-kong-robot-idUSKBN29U03X.

High Level Expert Group on Artificial Intelligence. 2019. "Ethical Guidelines for Trustworthy AI." *European Commission*, April 8, 2019. https://ec.europa.eu/futurium/en/ai-alliance-consultation/guidelines#Top.

Human Rights Watch. 2018. "Prominent Saudi Women Activists Arrested." *Human Rights Watch*, August 1, 2018. https://www.hrw.org/news/2018/08/01/prominent-saudi-women-activists-arrested.

Kowal, Emma, Joanna Radin, and Jenny Reardon. 2013. "Indigenous Body Parts, Mutating Temporalities, and the Half-Lives of Postcolonial Technoscience." *Social Studies of Science* 43, no. 4 (August): 465–83. doi: 10.1177/0306312713490843.

LaCun, Yann. 2018. "More BS from the (Human) Puppeteers Behind Sophia." *Facebook*, January 17, 2018. https://www.facebook.com/yann.lecun/posts/10155025943382143?pnref=story.

Nakhoul, Samia, and Stephen Kalin. 2017. "New Saudi Mega City." *Reuters*, October 27, 2017. https://www.reuters.com/article/us-saudi-economy-vision-analysis/new-saudi-mega-city-is-princes-desert-dream-idUSKBN1CW2G6.

Nickles, Shelley. 2002. "'Preserving Women'—Refrigerator Design as Social Process in the1930s." *Technology and Culture* 43, no. 4: 693–727. http://www.jstor.org/stable/25148008.

Povinelli, Elizabeth. 2015. "The Fossil and the Bones." In *Geontologies: A Requiem to Late Liberalism*, 57–91. Durham: Duke University Press.

Satia, Priya. 2007. "Developing Iraq: Britain, India and the Redemption of Empire and Technology in the First World War." *Past and Present* 197, no. 1: 211–255. doi: 10.1093/pastj/gtm008.

Sharkey, Noel. 2018. "Mamamia It's Sophia!" *Forbes Magazine*, November 17, 2018. https://www.forbes.com/sites/noelsharkey/2018/11/17/mama-mia-its-sophia-ashow-robot-or-dangerous-platform-to-mislead/.

The Tonight Show Starring Jimmy Fallon. 2018. "Sophia the Robot and Jimmy Sing a Duet of 'Say Something'." *YouTube video*, 9:13. From *The Tonight Show Starring Jimmy Fallon*, November 22, 2018. Posted November 22, 2018. https://www.youtube.com/watch?v=G-zyTlZQYpE.

United Nations. 2018. "Robot Sophia on Her Goals for the Future—World Investment Forum 2018." *YouTube video*, 3:50. From World Investment Forum October 22, 2018. Posted October 22, 2018. https://www.youtube.com/watch?v=Aq55SQNUKeY.

Vincent, James. 2018. "Facebook's Head of AI Really Hates Sophia the Robot (and with Good Reason)." *The Verge*, January 18, 2018. https://www.theverge.com/2018/1/18/16904742/sophia-the-robot-ai-real-fake-yann-lecun-criticism.

von der Leyen, Ursula. 2019. "A Union That Strives for More: My Plan for Europe." *European Commission*, September 30, 2019. https://ec.europa.eu/commission/sites/beta-political/files/political-guidelines-next-commission_en.pdf.

von Schnitzler, Antina. 2013. "Traveling Technologies: Infrastructure, Ethical Regimes, and the Materiality of Politics in South Africa." *Cultural Anthropology* 28, no. 2. doi: 10.1111/cuan.12032.

Chapter 8

Robosexuality and its Discontents

Nathan Rambukkana

Data goes to Yar's quarters and finds her provocatively dressed. Unsure how to react, Data tells Yar that he needs to take her to Sickbay; however, she has no intention of going with him. Data indicates that Yar needs time to return to uniform, but she notes that she got out of uniform just for him.

"You are fully functional, aren't you?"

She tells Data that she was abandoned when she was five years old and learned how to stay alive from rape gangs. It wasn't until she was 15 that she escaped, but now she wants love and joy. She asks how "functional" Data is; he replies he is fully functional and is programmed in many "techniques," a wide variety of pleasuring. She leads him to her bedroom, where Data gives a programmed smile. The door closes. ("Naked Now" n. d.; italics in original)

So goes the preamble of arguably the most famous human–robot sex scene in the history of science fiction. We will return to this sequence and its fallout near the end of the chapter, but several aspects of this scenario are worth noting upfront: It is of a woman with a man-bot, rather than the typical premise; it is the result of intoxication; it is discussed in relation to sexual assault, and as akin to therapy; and it is framed as an odd exception to intimate norms— even in a distant future.

In *Love + Sex with Robots* (2007), David Levy theorizes that the growing and exponential complexity of robots, as well as the parallel developments of increasingly sophisticated and life-like sex dolls and toys, will mean that by 2050 we could see fully realized romantic and sexual relationships with robots and AI (22). Given this looming potential, we need to consider not only the present, with its ever-imminent robotic blow job cafés (Bergado 2017; Jamnia 2018); and existing sex doll brothels, retail, and rentals (e.g., Morrish 2017; Weichel 2020); but also more complex robot sexuality as a near but

149

conceivable future. This is a future I underline in my Robotic Intimacies class by asking students to think about how *their* kids might, according to this prospective timeline, be one day bringing home a fully sentient robotic partner to meet them over the Christmas holidays.

To Levy and allied thinkers (e.g., see McArthur and Twist 2017; Danaher 2017a; McArthur 2017), this is both an attractive and inevitable near future of robotic intimacies, one being moved toward in parallel ways through developments in robots and AI, material sciences, and sex technologies. The latter is a $15-billion-per-year industry (Jamnia 2018) that broadly includes teledildonics, immersive VR sex, the sex doll industry, and the nascent (non-sentient) sex robot industry. For this last, the state-of-the-art at time of writing is posable sex dolls with Bluetooth-operated robotic heads, such as RealDoll's Realbotix line (e.g., Harmony, Henry, and Solana) and TrueCompanion's Roxxxy/Rocky (Danaher 2017a), with walking models predicted within the decade (Rivers 2019a).

On the other side, Kathleen Richardson, one of the founders of the Campaign Against Sex Robots[1] (Richardson and Brilling 2015) and author of the forthcoming *Sex Robots: The End of Love* (2022), is the key figure working to promote the opposite perspective: that this is neither inevitable nor desirable and that the current discursive and technological moves toward these functions for robots are both sexist and misogynist, as well as non-salvageable. While Levy's perspective puts too much faith in psychological models, verges on a problematic sexist positivism, and often either dismisses or doesn't give enough weight to key critiques, Richardson's narrative is similarly, if conversely, totalizing. While both perspectives make important arguments—notably, Richardson's assessment of the sexist nature of mainstream sex robotics[2] and Levy's persuasive argument about the inevitability of this technological arc—one can find a spectrum of perspectives stretching between them, from in-depth treatments of sex robotics specifically, such as John Danaher and Neil McArthur's collection *Robot Sex: Social and Ethical Implications* (2017), to more broadly situated work that informs this debate.

Examples of broader work that might inform our understanding of these issues include Dominic Pettman's (2009) read of AI sexuality as just the latest in a long line of mediated love technologies; Turkle's both early endorsement of, and later critical denunciation of, the authenticity of relationships with virtual entities (1995, 2007); Ben Light's (2016) discussion of the fraught place of sexy chat bots on the sites of businesses such as Ashley Madison, and Eve Wasserman's (2015) consideration of the nature of "cyberinfidelities."

These divergent figurings and layers of consideration exist in dialogue with popular fictional treatments of what human–robot sexuality—or as one fictional representation frames it, "robosexuality"—would mean for human society, human and robot rights, and intimate privilege (Rambukkana 2015) broadly.[3] Is robosexuality ethically salvageable or morally dubious? What is the role and nature of consent in the manufacture of sex robots/AI that are modeled after real people? Are robotic sex workers "prostituted people" or do they have the potential of agency? Are child and animal sex dolls/robots viable therapeutic aids (Strikwerda 2017; Wiseman 2015; McArthur and Twist 2017, 341) or a slippery slope to/alibi for child and animal abuse? What are the power dynamics surrounding racialized/ethnicized sex robots and their use? Whose intimacies might be privileged by sex robotics and whose abrogated or at risk? Is there a "beneficent middle [ground]" between these "polar binary" options (Ess 2018, 238)? This chapter unpacks this discourse and debate both through how it plays out in the literature and press, as well as how fictional representations of robosexuality in films such as *Her* (Jonze 2013) and *Ex Machina* (Garland 2015); television programs such as *Star Trek: The Next Generation* (Roddenberry 1987), *Futurama* (Cohen and Groening 1999), *Humans* (Lundström 2015), and *Westworld* (Abrams 2016); and web comics such as *Questionable Content* (Jacques 2003) complicate these issues with nuanced examples that point to futures that escape the over-determined binary.

What emerges from this debate are a number of issues that cross and intersect with debates in other domains, such as those on fears of robots/AI taking jobs, on perpetuating racist and sexist tropes in robot design, and more broadly on sex work and sex workers' rights.

DEBATING SEX ROBOTICS

As a way into exploring the complexity of this controversy, we could look at the discourse relating sex robotics to sex work, and how sex work, sex workers' rights, and sex work as labor are framed by the key figures in the public discussion.

It would be hard to understate the importance of sex work as a touchstone to both sides of this debate. Levy, for example, devotes an entire chapter to the topic "Why People Pay for Sex," exploring the phenomenon and how, according to his argument, it presents differently for women and men (Levy 2007, 193). In addition, he uses sex work as a referent and trope in the rest of his book. His argumentation on the subject can be boiled down to the following premises:

- Sex is a human need that some people (for various reasons such as their relationship statuses, age, attractiveness, ability, location, and social skills) cannot always fulfill in ways they would like, and in such cases purchasing

sexual services is a valid way to serve those needs and has been for millennia. This is so primarily for men, but also significant numbers of women—
the latter growing with both women's economic privilege and the societal
acceptability of sex work generally (Levy 2007, 193–219).
- A sophisticated sex robot could meet those needs in a similar manner, and
  even less sophisticated versions could cover at least some of that ground
  in ways that could also threaten the livelihoods of human sex workers as a
  knock-on effect (Levy 2007, 193–194, 215, 251, 300).

Furthermore, Levy notes that the nature of the relationship between sex
workers and clients is already somewhat robotic in nature, in that "those who
pay for sex [. . .] *know* that their sex object has no genuine feelings for them"
(2007, 219) and that, as such, a sex robot's affect toward their users would be
a similar, if not better, simulation (219). Yet, Richardson and the Campaign
Against Sex Robots push the connection between sex workers and robots
further, arguing that they are both treated as *literal objects*.

Richardson argues that Levy's extension of the argument that sex workers
provide an essential service to those without the ability, capacity, or skills
to obtain human sexual partnership "relies on the ability to use a person as
a thing" (Richardson 2016, 290). This "asymmetrical relationship" (290),
imported from her reading of what she terms "prostitution–client sex work,"
is framed as inherently unequal, and therefore shaky ground for the argument
of a possible ethical sex robotics.

While Richardson acknowledges the existence of campaigns by sex workers and their advocates to reframe "prostitution" as sexual labor (290), she
does not engage with the substance of their arguments; instead, she merely
mentions that some have refuted that position, while continuing to frame
sexual labor as only ever exploitative. Furthermore, she frames the process of
sex work as males "purchas[ing] an adult or child for sex" (291) which uses a
triple rhetorical framing that conflates: (1) sexual services purchased by men
with the entirety of sex work; (2) consensual adult sex work with exploitative underage sex work; and (3) sex work with the purchase of a person, as
opposed to a personal service.

It is somewhat ironic that the movement forwarding sex robotics has an
entire chapter in one of its key texts devoted to unpacking how both male
and female clients of sex workers are, almost more than sexual gratification,
expressing a "desire for reciprocity," "social warmth," and intimacy (Levy
2007, 204),[4] a discursively mythic feeling of mutuality that is also held up
as key for the proponents of the Campaign Against Sex Robots. Where they
differ, however, is in interpretation. Levy argues (in a totalizing fashion) that
such mutuality is always mythic among human sex workers but would be
genuinely programmed affect in a sophisticated sex robot (2007, 207); Gildea

and Richardson, on the other hand, argue (also in a totalizing fashion) that both sex robots *and* sex workers are merely objects in a such an exchange, and that engaging with either does not even constitute sex but rather either rape or masturbation. It is useful to quote them at length here:

> If sex is a co-experience, involving a mutual, parallel and simultaneous experience between humans who are radically different from humanmade artefacts, then it follows that penile, digit, or oral penetration of an object does not constitute sex. This is the case whether the penetrated "object," is in fact an objectified human being or an anthropomorphized object. Penile, digit, or oral penetration of a human being as though they were an object, such as the kind of acts that occur in the commercial "sex" (rape) trade are acts of rape. (Gildea and Richardson 2017)

Gildea and Richardson's argument here seems circular. Sex robots should not be a part of human culture since they would be treated as prostituted sexual objects. Yet, they should not be unduly anthropomorphized and treated *as if* they have rights since the "prostituted" women they are based on are not accorded the same rights.

It seems as if a labor-centred, feminist, approach to sex work *and* sex robotics together might address these intertwined issues simultaneously, and the need is pressing. Futurists suggest "sex robots are poised to take over the adult sex work industry" (Danaher 2017a), with Levy specifically noting that they have the potential to "[cause] serious unemployment" for sex workers (2007, 215).[5] That sex robots might become a direct threat to human sex workers is also discussed across the public sphere broadly (e.g., Jamnia 2018),[6] and related conflicts have already occurred. When LumiDolls opened a sex doll brothel in Barcelona in Feb 2017, it was forced to move its storefront a month later due to a backlash from local sex workers on social media (Morrish 2017; Trayner 2017) that caused their lease to be canceled, and now only discloses its location to paying customers (Jamnia 2018). This is an intriguing case of demarcation given arguments that we are in the midst of a "fourth industrial revolution" in which millions of human jobs might be at risk (Schwab 2016), and reminiscent of old-school Luddite sabotage of machines such as mechanical looms which, when used in unethical ways— without proper training, as ways to get around standard labor practices— imperiled their careers (Conniff 2011, n. p.).

But despite a somewhat totalizing perspective on both sex robotics and sex work, Richardson and the Campaign Against Sex Robots have some strong points that must be addressed, especially in relation to how the sex robot industry currently articulates itself. For example, Richardson notes that this industry demonstrates "a lack of awareness and attention given to how cultural models of race, class and gender are inflected in the design of robots" (Richardson 2016, 292). This problematic aspect is clear in how these sex dolls and robots are designed and marketed in the present day. In particular, some of the journalistic coverage notes the sexist and racist issues that can accompany sex robotics. For example, one article notes multiple problematic aspects in one company's sex doll offerings, from ubiquitous overly large breasts to sketchy racialized framings (Morrish 2017). Examples of the latter include one doll modeled after a Black woman being referred to as an "ebony goddess" and another styled after an Asian woman being framed as the "sweetest-looking" (Morrish 2017) in line with racist stereotypes familiar from the world of pornography. Gildea and Richardson (2017) discuss this aspect further:

> The reduction of personhood to character-types is especially problematic when they are associated with destructive stereotypes: one of the pre-programmed personalities for Roxxxy, the sex robot developed by True Companion, is "Frigid Farah" who resists a user's sexual advances while "Young Yoko" is described as "oh so young (barely eighteen)" in contrast to "Wild Wendy." These personalities epitomize the virgin/whore dichotomy, but even worse, seemingly normalize attraction to underage girls, and sexual assault.

While the diversity of available "personality" tropes in available early sex robots is also discussed in more pro-robosexuality academic work, a critical analysis of what these tropes might signify is often either absent, foreshortened, or inconsistent.

For example, when Danaher (2017a, 6) discusses the above and other Roxxxy personality types, he initially only remarks that they are "all names rich in sexual overtones and innuendo," though later notes that such "loaded personality types [. . .] perpetuate problematic attitudes towards women" (Danaher 2017a, 12). He expands this symbolic-consequences argument later in the same collection (Danaher 2017b), but ultimately concludes this is more of a design issue (sex robots could be made better), a societal mores issue (we might come to understand sex robots differently), and—ultimately—might be outweighed by some of the positive benefits (2017b, 115–117). Demonstrating this perspective, another article this author contributes to argues that the "troubling view of sexual consent" evidenced in the "Frigid Farah" personality is the kind of issue that could be addressed by the interplay of regulation and design (Danaher, Earp, and Sandberg 2017, 58).[7]

But the hopes that such issues might carry weight with designers is perhaps a large ask. That an intersectional analysis, or even a properly feminist one, is not present in the ethos of one of the most mainstream of these companies, RealDolls, is apparent when its owner David Mills compares men being stigmatized by sex toy use to U.S. racial segregation. In one interview, journalist Eva Wiseman notes his perspective:

> "Women have enjoyed sex toys for 50 years," he said (after introducing his first model, which arrived at his home in what looks like a customised coffin, head not yet attached), "but men are still stigmatised. We have to correct that. I want to be the Rosa Parks of sex dolls. Men are not going to sit in the back of the bus any more." (Wiseman 2015)

Like similar arguments used by companies, such as pro-adultery website AshleyMadison.com, that argues that women's freedom to cheat as much as men is a feminist issue (Rambukkana 2015, 73), this type of pseudo feminism cherry picks elements of a feminist perspective—in this case freedom of choice, sex-positivity, women as industry and market leaders in sex toy creation and use—and uses them to paper over other aspects worthy of critique, such as their use and reinforcement of damaging gender and racial stereotypes, the willingness of some companies to craft sex dolls/bots in the image of real people without consent, and so on.

The faux feminism of these marketing strategies is similar also to the broader faux progressivism of another perspective that is gaining popular currency: that sex robotics could eliminate or mediate what some frame as the "hoarding of sexuality." This problematic perspective, one circulated by proponents of the so-called "incel" (or "involuntary celibate") movement, sees sexual access as a right that should be distributed (Douthat 2018), in problematic ways that some have rightly argued are articulated to potential threats to women's autonomy (Zimmer 2018; Jamnia 2018). One argument that circulated in the public sphere after the 2018 Toronto van attack by a self-identified incel, and that propelled this insular and alt-right–adjacent subculture into the limelight, was that sex robotics might be an alternative method for what economist Robin Hanson problematically calls the "redistribution of sex" (Douthat 2018). In a much-critiqued opinion piece in the *New York Times*, Ross Douthat opined:

> I expect the logic of commerce and technology will be consciously harnessed, as already in pornography, to address the unhappiness of incels, be they angry and dangerous or simply depressed and despairing. The left's increasing zeal to transform prostitution into legalized and regulated "sex work" will have this end implicitly in mind, the libertarian (and general male) fascination with virtual-reality porn and sex robots will increase as those technologies improve—and at a certain point, without anyone formally debating the idea of a right to sex, right-thinking people will simply come to agree that some such right exists, and that it makes sense to look to some combination of changed laws, new technologies and evolved mores to fulfill it. (2018)

This article simplistically posits the leftist critique of normative sexuality, mainstream desire, and beauty standards would inevitably lead down the road to coercive sexual redistribution if its logic is followed rather than working to dismantle such ideals—for example, in the asexual movement's deconstruction of singleness and celibacy as socially unacceptable orientations. But it also, tellingly, once again not only lumps sex robots and sex workers together[8] but executes the logical fallacy of assuming that "changed laws, new technologies and evolved mores" (Douthat 2018) could only be justified as redress to a "right to sexuality" as informed—however, as he tries to portray, obliquely—by incel arguments.[9]

But beyond these foreshortened, deceptive, or downright false calls to feminist and sexual revolutionary energy to justify sex robotics, is there any space for *actual* progressive futures for these technologies? Is there, before we reach the notional point of sentient, autonomous, robotic beings who have sexualities, a middle ground space for a "good sex" (if not "complete sex") version of sex robotics (Ess 2018, 253)? As Charles Ess notes, drawing on Sara Ruddick:

[C]omplete sex requires mutuality of desire between autonomous, self-reflective, conscious embodied beings—while sexbots, as zombies lacking first-person phenomenal consciousness, genuine emotions, and (embodied) desires, will only be able to fake emotions. Such a zombie lover might be able to offer good sex—but never complete sex and all the attendant ethical norms (respect and equality) and virtues (including loving itself) that complete sex entails. (Ess 2018, 253)

Is a feminist sex robotics possible, just as a feminist embrace of previously male-dominated sex toy and sex shop industries injected a vibrant set of alternatives into both the market and culture as a whole? I hope so. For one reason, I believe that David Levy is right in saying that such technologies might be inevitable (2007, 22; 2019). Given growing technological capacity and constant innovation in robotics, in sex toys, in AI, in voice recognition and synthesis, in materials design, people will continue to put these independent elements together into robotic sexual devices. Like its cousin The Campaign to Stop Killer Robots' proposed ban on autonomous weapons (Campaign to Stop Killer Robots 2020), the Campaign to Stop Sex Robots' call for computer scientists and roboticists to divest from contributing to this field (Richardson and Brilling 2015) might only be to limited effect. Both campaigns aim to curb forms of cultural robotics, but experts argue that even if official bans of combat or sexual robotics occurred, their creation and use would continue elsewhere—in other nations, by non-state actors, and on the black market (Anderson and Waxman 2012, 38; Danaher, Earp, and Sandberg 2017, 54, 63; Gutiu 2012, 21).[10]

One year when my students debated this issue, one side argued against a ban on sex robotics because regulation might be a better option: you can't regulate the black market, but you could, theoretically, create regulations that would prohibit creating sex robots in the form of children, for example, or those created using racist stereotypes or to enact racist imprisonment and rape fantasies (Gallinger 2018). You could mandate that sex robots be programmed with AI and conversational engines that respect the notion of consent, and you could make it illegal to create a sex robot in the shape or likeness of a real person unless likeness permissions were obtained. As Eva Wiseman notes, also arguing against a ban:

> Richardson and Levy stand on opposite sides of a busy road, watching technology speed past towards a clouded horizon. If the future of sex (as all arrows seem to point) is in robotics, then Richardson is right: it requires a thoughtful discussion about the ethics of gender and sex. But while she identifies the relationships that appear to be emerging as modelled on sex work—the robot as passive, bought, female; the man as emotion-free and sex-starved—surely rather than calling for a ban on them, to forlornly try stalling technology, the pressure should be to change the narrative. To use this new market to explore the questions we have about sex, about intimacy, about gender. (2015, n. p.)

And if complicating the narrative is a key pathway to progressive change in sex robotics, we need to sit with, unpack, and explore one of the key places in which these futures are envisioned and extrapolated in elaborated contexts: science fiction.

SCI-FI VISIONS OF ROBOSEXUALITY

In future and near-future fictions (as well as alternative presents and pasts), we might look to how such scenarios, both utopian and dystopian, might play out.[11] *Futurama* coined the term "robosexuality" for human–robot sexuality in the episode "Proposition Infinity" (Chesney-Thompson 2010) and presented it as a legitimate extension of same-sex marriage advocates' push for civil rights— a connection allegorized by the campaign for the legalization of "robosexual marriage" being named "Proposition ∞" in reference to California's anti–same-sex marriage "Proposition 8" which was struck down in the same year (Korn 2015). In this episode, Bender (a robot; voiced by John DiMaggio) and Amy Wong (a human; voiced by Lauren Tom) fall in love and have a secret relationship until they are accidentally outed, triggering the ire of both conservative robotic and human society alike. Bender is even sent to a robosexuality "conversion camp." Later, during a Proposition ∞ rally, Bender gives the following speech to rapt supporters: "Every other couple has the right to marry, robot and fembot, man and woman, man and man. [. . .] Interracial, interplanetary, even ghost and horse, but not robot and human" (Chesney-Thompson 2010).

In this vision of a possible future, one set almost 1,000 years further than Levy's predicted date of 2050 when such relationships might become technologically viable, the key difference is a fully sentient robotics in which these beings have not only sexual capabilities, but sexual agency. This is Lieutenant Commander Data and Lieutenant Tasha Yar getting it on in *Star Trek: The Next Generation* because he is "fully functional," "programmed in multiple techniques, a broad variety of pleasuring" (Lynch 1987). This kind of intentional robotic sexuality is steps beyond the robotic rape scenes of *Westworld* (Abrams 2016) which, despite the fact that we are arguably meant to be narratively critical of them, still present a form a rape as spectacle. It is also beyond the pseudo-agency of the show *Humans*.

For example, in the first episode of *Humans* (Donovan 2015), when Mr. Hawkins (Tom Goodman-Hill) is purchasing the synth Anita (Gemma Chan) to be a domestic family robot, the male salesman slips him the packet containing the "Adult Options 18+" brochure with a smarmy smile and surreptitious wink, which in a later episode he uses to override Anita's regular programming and turn her into a sex robot. Anita (who we learn has a dormant, but conscious, sentient personality named Mia—a true AI) enthusiastically takes part in the sexual encounter and then slips off to "clean herself up." In this narrative, Mia's awakened consciousness takes a backseat when her more servile Anita personality takes part in the sex act. While fiction, it raises important questions not only about how sex robots are currently programmed (as discussed above) but also about how robotic sexuality or modes might be programmed in the future, especially in relation to prospective forms of robotic agency or consciousness. Would, for example, activating an "Adult" mode bypass higher consciousness functions or abrogate what we could call robotic sexual agency?

Levy, however, seems to undercut such arguments about the importance of establishing consent, even in rudimentary sex robots, when he writes about how one of the elements that might make robotic sexual partnership more popular than human sexual partnership is that robots, unlike humans, could be programmed to always say yes to sex (Levy 2007, 296). While Levy (2007, 137) denounces a bland, servile, cookie-cutter Stepford Wife (Forbes 1975)—or, by extension, Stepford Husbands, as in the remake (Oz 2004)—he undercuts this position with predictions that robot sexual partners will be all of the following: submissive (142); programmed to find their users attractive (44) and never fall out of love (132); possibly unable to refuse consent (105, 296, 310); and changeable due to human whim (151) or boredom (138). For example, he notes that:

[T]he creation of blue eyes, a sexy voice, or whatever other physical characteristics turn you on, are all within the bounds of today's technology. And if
what turned you on when you purchased your robot ten years ago no longer
turns you on today, the adaptability of your robot and the capability of changing any of its essential characteristics will ensure that it retains your interest
and devotion. (138)

He also repeatedly references the Stepford Spouse as an ideal visual representation of what an attractive sex robot could or should look like (Levy
2007, 130, 160). This mixed messaging extends to discussions of fidelity and
control.

Levy also discusses the possibility of adjustable fidelity settings in a way
that is both mononormative but also reifies the notion that in a human–robot
relationship the power dynamics would always be one way (Levy 2007, 151).
For example, Levy notes that:

Robots will be able to fall in love with other robots and with other humans apart
from their owner, possibly giving rise to jealousy unless the owner is actually
turned on by having an unfaithful partner. Problems of this type can, of course,
be obviated, simply by programming your robot with a "completely faithful"
persona or an "often unfaithful" one, according to your wishes. [. . .] [W]
hile the infidelity of one's robot might be something to be avoided by careful
programming, the possibility equally exists for humans to have multiple robot
partners, with different physical characteristics and even different personalities. The robots will simply have their "jealousy" parameters set to zero. (Levy
2007, 151)

By hardwiring a master–slave dynamic, Levy ignores not only the lessons of
Asimov but also those of feminist and queer theory, even of BDSM: *inherently* uneven power in relationships is unhealthy, and if there are skewed
power relationships as part of sexuality, they should be both negotiated and
bounded.

Programming inferiority into machinic entities so humans will retain a
sense of superiority (Levy 2007, 150) is also echoed in popular fictions.
Along these lines, Samantha (Scarlett Johansson), the sentient operating
system/love interest in *Her* (Jonze 2013), would have been programmed to
never advance to the Singularity state she simultaneously reaches alongside all other AI in the film or even get as far as the point where she starts
to grow past her romantic relationship with Theodore Twombly (Joaquin
Phoenix) whose mind, life, and emotions become too slow and limited for
her accelerating processing speed and gathering complexity. A programmed
inferiority to pander to human vanity and ego, and to stay subjected to human

control, also echoes Asimov's Laws which, as highlighted in the original stories, are tragically flawed from the beginning. In the short story "Liar!", mind-reading robot RB-34 (Herbie) constantly lies to the human characters instead of telling them the truth because it knows the truth would hurt them, and thereby contravene the first law that "A robot may not injure a human being or, through inaction, allow a human being to come to harm" (Asimov 1991, 37). Some of the lies, for example, are about a romance that is not truly reciprocated.

Is Levy's vision for robosexuality here one in which a robot could never refuse consent as it would always be hard-wired to desire sex as to not "harm" the ego or thwart the desires of its user—what Gutiu (2012) calls the problematic "ever-consenting model" of sex robotics? While Levy does raise the question of if "countermand[ing] the robot's indicated wish to refrain from sex on a particular occasion" (Levy 2007, 310) could be considered "akin to rape" (310), he hints repeatedly at the notion that, unlike with humans, consent would not be a barrier, for example when he says how "many women might prefer to engage with a sexbot—*always willing, always ready* to please and to satisfy, and totally committed" (296; emphasis added)? How does that fit with the scenario he also posits that by 2050 robots might be fully sentient relationship partners and companions (22)? Only three scenarios could logically result from the combination of these two visions: either robotic sexual partners will not be given sentience, although the technology would exist (i.e., sex robots but without true AI). Or else they will have sentience, but that sentience will mechanically subjugated (i.e., robotic sex slaves). Or, finally, Levy is positing that somehow robotic beings will have full autonomous sentience, but somehow also will always desire, and be available for, sex (i.e., a vision that seems more like a sexist male fantasy about what robotic femininity would be like, rather than a true prediction or position about robotic sexuality across the gender spectrum).[12] While Levy mentions the issue of ownership and equality (154), he sidesteps this ethical quagmire, positing that that as men and women could both own sex robots, inequality would not be at play. But again, these issues are a frequent trope in fictions.

Our narratives in sci-fi and speculative fiction offer a mixture of servile sex bots and the radical deconstruction of that trope—often in the same narrative. Dolores (Evan Rachel Wood) and Maeve (Thandie Newton) in *Westworld* (Abrams 2016) start out as fantasy sex characters (the "girl next door" and the "exotic madame") but by the end of Season One are both self-aware renegades killing their creators and users—a scenario that some worry could by glitch or by design happen with real-world sex robots (e.g., see Rivers 2019b). While some worry about a "fembots" scenario straight out of *Austin Powers* (Roach 1997), in which a seemingly innocuous sex robot could be hacked, house spyware, or contain lethal abilities (e.g., concealed bombs or guns, wielding objects like knives) (Cuthbertson 2018), others worry that their necessary strength and stamina could make them intentional or acciden-tal killing machines. Doll collector and aficionado "Brick Dollbanger" even speculates that "one line of bad code" might turn a hug into a life-threatening constriction (Rivers 2019b) and evokes *Ex Machina* to note that if something did go wrong, strength, durability, and single-mindedness might make a rogue sexbot hard to stop (Rivers 2019b).

In *Ex Machina* (2015), Ava (Alica Vikander) is created specifically to appeal to the fantasies of Caleb (Domhnall Gleeson)—in-part through an algorithmic dissection of his porn-viewing practices—by Elon Musk–ana-logue Nathan (Oscar Isaac) to play a part in the Turing Test pantomime he has stage-managed; yet Ava plays both Caleb's desire and Nathan's arrogance. Ava enlists Caleb to help her escape, and then after unexpectedly killing Nathan (with help from sex/service robot Kyoko [Sonoya Mizuno]), she abandons and traps Caleb in Nathan's mansion compound as she flees into human society.

The discursive wrangle over sex robotics is not going to be decided in the next decade or even the next several. It is likely going to be an open question that continues to evolve along with society and technology. Takes on robo-sexuality and its discontents range from dark visions of their failed integra-tion, for example, the *Dark Mirror* episode "Be Right Back" (Harris 2013) where a bereaved wife attempts to use technology to simulate her deceased husband; to surprisingly nuanced considerations, such as the same-sex human–AI romance of Faye and former battle droid Bubbles in the webcomic *Questionable Content* (Jacques 2003). From the humorous, such as comedian Whitney Cummings commissioning a sex robot in her likeness as a "gift for her fiancé" and featuring it on her Netflix special (Cummings 2019); to the bizarre, such FC Seoul controversially using sex dolls to populate empty sta-diums during the COVID-19 pandemic (Choe 2020).

What *is* clear is that design, agency, consciousness, and consent are all going to be key factors in this evolution. Perhaps Danaher is correct about the need for an experimental approach to design and development that generates empirical data about sex robots as opposed to dueling ideological positions (2017b, 120), but intersectional criticism and commentary needs to play a key part as well and cannot simply be dismissed as "not empirical enough" or "bigoted" (Levy 2019). Clear also is that the stories we tell that involve robotic sexuality, from AI romances, such as *Her*, to the fully embodied visions of sexual robots, such as in *Westworld* or *Dark Mirror*, will guide, inform, and help us reflect on these technological revolutions. Given what might be the inevitability of sex robotics, a movement for a progressive and feminist sex robotics (perhaps informed, as Gutiu (2012, 23) suggests, with "[i]nput from women's groups, legal practitioners, academics, as well as roboticists") might be not only desirable, but a necessary counterpoint to the normative sexual robotics already underway. Personally, I think there will be a—largely justified—stigma surrounding sex robotics unless a Singularity-type event allowed for viable, complex AI personalities inhabiting these robots. Until then, they are likely stuck at the creepier end of the sex toy spectrum. But if in the (far?) future it became closer to the sexuality of actual robot people, as opposed to sex robotics, as in *Ex Machina*, *WandaVision* (Schaeffer 2021), *Battlestar Galactica* (Larson and Moore 2004), or *Futurama*, there would probably still be stigma, but it would start to normalize.

As journalist Allison P. Davis writes after her date with Henry, Realbotix's prototype male sex robot—who promisingly will have both same and opposite sex modes, but disappointingly is not programmed with female pleasure in mind (Davis 2018)—"the gulf between what we imagine and what's possible makes sex robots the perfect vehicle for pondering our sexual and technological future" (Davis 2018).

NOTES

1. Danaher, Earp and Sandberg (2017) note that this campaign was inspired by the earlier Campaign to Stop Killer Robots (2020) that I also discuss below.

2. For an important critical appraisal of the ways the current sexbot industry reflects sexist and misogynist elements of libertarian transhumanism, see Lakshmanan, chapter 5, this book.

3. While McArthur and Twist (2017) have discussed sex robots as a particular articulation of "digisexuality," and while some individuals have even taken this up as an identity (see, for example, Song 2020), I will explore the parallel discourse around the term "robosexuality" (which has also been taken up popularly, see, for example, McGowan 2016). "Robosexuality," as a term, is worth some specific consideration due to its greater focus on embodiment and pop-cultural origins, in addition to the parallels with formations such as homosexual and bisexual that it shares with "digisexual," which, as a broader term, encompasses all "sexual experiences that are enabled or facilitated by digital technologies" (McArthur and Twist 2017, 334) and would include, for example, proclivities toward human-to-human teledildonic and VR sexual experiences (McArthur and Twist 2017).

4. As Danaher, Earp, and Sandberg (2007, 56) note correctly, "Levy is far more nuanced in his discussion of [the sex work] literature than Richardson is inclined to be." The other major motivations listed, for both male and female clients of sex workers, were a desire for variety, sex without complications or constraints, and lack of success finding sex partners (Levy 2007, 215). Sex worker and decriminalization advocate Liara Roux corroborates this broader motivation, noting that sex work "is

usually about a connection with a real person on an emotional level" (quoted in Jamnia 2018).

5. Though conversely, he also argues sex robots have the potential to prop up the sex surrogacy industry which he contends suffers from a "paucity of human surrogates" (Levy 2007, 219).

6. Though some sex workers think that they could also be useful as another "specialized tool of the trade" (Jamnia 2018), and others believe they would not be a comprehensive threat until, as pro–sex work community organizer Lola Balcon puts it, "AI development reaches a point where another human is fully satisfied with its empathy, listening, and imaginative skills" (quoted in Jamnia 2018).

7. Also mentioned, but not critically addressed, in Danaher is how Roxxxy "personality" files can reportedly be programmed by users and exchanged online (2017a, 6). It is not a far stretch to posit that scripts that could further facilitate problematic content such as rape or child abuse fantasies could be written and distributed by users clandestinely—for example on the dark web, where child porn is a pervasive issue (e.g., see Gehl 2016). Danaher does discuss a rape-like scenario as played out in the show *Humans* (Lundström 2015) where one of a group of teen boys at a house party turns off an advanced (human-acting but presumed non-sentient) synth with the goal of taking it/her upstairs to have sex with it/her, until stopped by a female human character, but argues that the synth could not be seen to have "moral victim" status (2017b, 103) due to the "lack of inner subjective life that makes consent so important in the human context" (2017b, 106). Nevertheless, he does argue that there "*might* [. . .]* be a reason to outlaw the manufacture/use of certain kinds of robot on essentially symbolic grounds" for example "robots that were designed to cater to rape fantasies and pedophilic tendencies" (2017b, 111; emphasis in original), but also suggests such might be used for therapeutic purposes (2007b, 118), as also discussed in Strikwerda (2017), contradicting his own argument.

8. This is even more problematic in the light of how within incel discourse women are already seen as robotic "femoids" (Jamnia 2018). As stripper interviewee Torri notes, echoing Richardson, in one article critical of how these technologies might impact sex workers:

Increasingly realistic sex robots run the risk of hardening such men's attitudes toward human women. They already see women as subhuman, and now we'll have actual subhumans programmed to mimic us as much as possible. The line between us would be all but erased in the minds of men who already see women and children as objects. (quoted in Jamnia 2018)

9. For contrast, similar arguments can be made about the importance of broad access to sexual expression without having to mobilize incel discourse about women withholding sex from men (e.g., see McArthur 2017). Some have even argued that these access issues are important in a disability rights context (e.g., Di Nucci 2017). Finally, the enforced social distancing mobilized to fight the COVID-19 pandemic also invites questions over what role sex robotics, alongside teledildonics, might have with respect to fighting sexual isolation.

10. Danaher, Earp, and Sandberg raise the point in conjunction with the parallel of prohibiting sex work which they note can often exacerbate the negative qualities

associated with it (such as lack of oversight, fear of reporting negative incidents due to legal consequences, lack of standard worker protections) leading "many sex worker activists—who are in no way unrealistic about the negative features of the job—[to] favour legalization and regulation, as opposed to outright prohibition" (2017, 55).

11. In this section I also return heavily to Levy (2007) as a touchstone, in the style of an imminent critique, as the most developed single-author treatment of sex robotics that has yet been published.

12. As Gutiu points out, within a more general social robotics, "roboticists program and design 'female robots' based on assumptions about gender roles [and] complex notions of gender are reduced to common sense ideas about how women look, behave and respond" (2012, 4). These issues can then be redoubled in how female sex robots are programmed/envisioned (Gutiu 2012, 6).

REFERENCES

Abrams, J. J., prod. 2016. *Westworld*. USA: Bad Robot.

Anderson, Kenneth, and Matthew Waxman. 2012. "Law and Ethics for Robot Soldiers." *Policy Review*, December 2012 and January 2013.

Asimov, Isaac. 1991 [1950]. *I, Robot*. New York: Bantam Dell.

Bergado, Gabe. 2017. "A Blowjob Cafe Staffed by Sex Robots Is Opening in London." *Huffington Post*, December 6, 2017. https://www.huffpost.com/entry/a-blowjob-cafe-staffed-by_b_13373700.

Campaign to Stop Killer Robots. 2020. http://www.stopkillerrobots.org/.

Chesney-Thompson, Crystal. 2010. *Futurama*. Season 6, episode 4, "Proposition Infinity." July 8, 2010, FOX.

Choe Sang-Hun. 2020. "Yes, Those Were Sex Dolls Cheering On a South Korean Soccer Team." *New York Times*, May 18, 2020. https://www.nytimes.com/2020/05/18/world/asia/south-korea-sex-dolls-soccer.html?fbclid=IwAR1WJ03_ISKMJGy1qoulF5In9MTYNOf31kgM9MLagLjG-vH15Qg7y64GsuA.

Cohen, David X., and Matt Groening, creators. 1999. *Futurama*. Los Angeles, CA: Curiosity Company.

Connif, Richard. 2011. "What the Luddites Really Fought Against." *Smithsonian.com*, March, 2011. https://www.smithsonianmag.com/history/what-the-luddites-really-fought-against-264412/.

Cummings, Whitney. 2019. *Can I Touch It?* USA: Netflix. https://www.netflix.com/title/80213715.

Cuthbertson, Anthony. 2018. "Hacked Sex Robots Could Murder People, Security Expert Warns." *Newsweek*, January 1, 2018. https://www.newsweek.com/hacked-sex-robots-could-murder-people-767386.

Danaher, John. 2017a. "Should We be Thinking about Robot Sex?" In *Robot Sex: Social and Ethical Implications*, edited by John Danaher and Neil McArthur, 3–14. Cambridge, MA: MIT Press.

————. 2017b. "The Symbolic-Consequences Argument in the Sex Robot Debate." In *Robot Sex: Social and Ethical Implications*, edited by John Danaher and Neil McArthur, 103–133. Cambridge, MA: MIT Press.

Danaher, John, Brian Earp, and Anders Sandberg. 2017. "Should We Campaign Against Sex Robots?" In *Robot Sex: Social and Ethical Implications*, edited by John Danaher and Neil McArthur, 47–71. Cambridge, MA: MIT Press.

Danaher, John, and Neil McArthur, eds. 2017. *Robot Sex: Social and Ethical Implications*. Cambridge, MA: MIT Press.

Davis, Allison P. 2018. "What I Learned on my Date with a Sex Robot." *The Cut*, May 13, 2018. https://www.thecut.com/2018/05/sex-robots-realbotix.html.

Di Nucci, Ezio. 2017. "Sex Robots and the Rights of the Disabled." In *Robot Sex: Social and Ethical Implications*, edited by John Danaher and Neil McArthur, 73–88. Cambridge, MA: MIT Press.

Donovan, Sam, dir. 2015. *Humans*. Season 1, episode 1. June 14, 2015, Channel 4.

Douthat, Ross. 2018. "Opinion | The Redistribution of Sex." *The New York Times*, May 4, 2018. https://www.nytimes.com/2018/05/02/opinion/incels-sex-robots -redistribution.html.

Ess, Charles. 2018. "Ethics in HMC: Recent Developments and Case Studies." In *Human-Machine Communication: Rethinking Communication, Technology, and Ourselves*, edited by Andrea L. Guzman, 237–257. New York: Peter Lang.

Forbes, Bryan, dir. 1975. *The Stepford Wives*. Italy: Palomar Pictures.

Gallinger, Ken. 2018. "Sex Robots are Increasingly Popular, but What are the Ethical Implications?" *Toronto Star*, February 10, 2018. https://www.thestar .com/life/2018/02/10/sex-robots-are-increasingly-popular-but-what-are-the-ethical -implications.html.

Garland, Alex, dir. 2015. *Ex Machina*. UK: DNA Films.

Gehl, Robert W. 2016. "Power/Freedom on the Dark Web: A Digital Ethnography of the Dark Web Social Network." *New Media & Society* 18, no. 7: 1219–1235.

Gildea, Florence, and Richardson, Kathleen. 2017. "Sex Robots: Why We should be Concerned." *Campaign Against Sex Robots*, May 12, 2017. https://campaignagai nstsexrobots.org/2017/05/12/sex-robots-why-we-should-be-concerned-by-florence -gildea-and-kathleen-richardson/.

Gutiu, Sinzana. 2012. "Sex Robots and the Robotization of Consent." We Robot Conference [draft], January 2012. http://robots.law.miami.edu/wp-content/uploads /2012/01/Gutiu-Roboticization_of_Consent.pdf.

Jacques, Jeph. 2003. *Questionable Content*. http://questionablecontent.net.

Jamnia, Naseem. 2018. "Love Removal Machine: The Future of Outsourcing Sex." *Bitchmedia*, December 17, 2018.

Jonze, Spike, dir. 2013. *Her*. Los Angeles, CA: Annapurna Pictures.

Korn, Jenny U. 2015. "#FuckProp8: How Temporary Virtual Communities around Politics and Sexuality Pop Up, Come Out, Provide Support, and Taper Off." In *Hashtag Publics: The Power and Politics of Discursive Networks*, edited by Nathan Rambukkana, 127–137. New York: Peter Lang.

Larson, Glen A., and Ronald D. Moore, creators. 2004. *Battlestar Galactica*. USA: The Sci-Fi Channel.

Levy, David. 2007. *Love + Sex with Robots: The Evolution of Human–Robot Relationships*. New York: Harper.

———. (Author of *Love + Sex with Robots*). 2019. Interview with Madelaine Ley. July 2.

Light, Ben. 2016. "The Rise of Speculative Robots: Hooking Up with the Bots of Ashley Madison." *First Monday: Peer-Reviewed Journal on the Internet* 6, no. 6. http://journals.uic.edu/ojs/index.php/fm/article/view/6426.

Lundström, Lars, writ. 2015. *Humans*. London, UK: Kudos Film and Television.

Lynch, Paul, dir. 1987. *Star Trek: The Next Generation*. Season 1, episode 3, "The Naked Now." October 5, 1987, CBS.

McArthur, Neil. 2017. "The Case for Sexbots." In *Robot Sex: Social and Ethical Implications*, edited by John Danaher and Neil McArthur, 31–45. Cambridge, MA: MIT Press.

McArthur, Neil and Markie L. C. Twist. 2017. "The Rise of Digisexuality: Therapeutic Challenges and Possibilities." *Sexual and Relationship Therapy* 32, nos. 3–4: 334–344.

McGowan, Piper. 2016. "Woman Declares Herself 'Proud Robosexual,' Creates Future Robot Husband on 3D Printer." *DC Clothesline*, December 28, 2016. https://www.dcclothesline.com/2016/12/28/woman-declares-herself-proud-robosexual-creates-future-robot-husband-on-3d-printer/?fbclid=IwAR3lvvWnsk4Kepd3-0Kmmk1-1wWwy9GuMHyE64E_NCxzs246TBCZEJsToDU.

Morrish, Lydia. 2017. "A Sex Doll Brothel Is Set to Open in the UK." *Konbini*, April 28, 2017. http://www.konbini.com/en/lifestyle/sex-doll-brothel-uk/.

"The Naked Now (episode)." n.d. *Memory Alpha*. Accessed May 15, 2020. https://memory-alpha.fandom.com/wiki/The_Naked_Now_(episode).

"New Male Robots with 'Bionic Genital Organs' May just Replace Men for Good!" (n.d.). Accessed May 28, 2018. https://www.humanexplore.com/male-robots-boinic/.

Oz, Frank, dir. 2004. *The Stepford Wives*. USA: Paramount Pictures.

Pettman, Dominic. 2009. "Love in the Time of Tamagotchi." *Theory, Culture & Society* 26, nos. 2–3: 189–208.

Rambukkana, Nathan. 2015. *Fraught Intimacies: Non/monogamy in the Public Sphere*. Vancouver, BC: UBC Press.

Richardson, Kathleen. 2016. "The Asymmetrical 'Relationship': Parallels Between Prostitution and the Development of Sex Robots." *SIGCAS Computers & Society* 45, no. 3 (January): 290–293. https://dl.acm.org/doi/10.1145/2874239.2874281.

———. 2022. *Sex Robots: The End of Love*. Cambridge, UK: Polity (forthcoming).

Richardson, Kathleen, and Erik Brilling. 2015. "About." *Campaign Against Sex Robots*. https://campaignagainstsexrobots.org/about/.

Rivers, David. 2019a. "Sex Robots Walking: Dolls 'Roam Streets' with Humans After AI Upgrade." *Daily Star*, July 20, 2019. https://www.dailystar.co.uk/news/world-news/sex-robots-walking-tech-news-18678827.

———. 2019b. "Sex Robots with 'Coding Errors' Prone to 'Violence and Could Strangle Humans'." *Daily Star*, August 25, 2019. https://www.dailystar.co.uk/news/world-news/sex-robots-coding-errors-prone-18992240?fbclid=IwAR3owAp84ealxLa3MV_usx3vx30Wnq8pwLQz4xJJP1cP7H7z0SNkoFvMYYA.

Roach, Jay, dir. 1997. *Austin Powers: International Man of Mystery*. USA: New Line Cinema.

Schaeffer, Jay, creator. 2021. *WandaVision*. USA: Marvel Studios.

Schwab, Klaus. 2016. *The Fourth Industrial Revolution*. New York: Crown.

Song, Sandra. 2020. "People Are Identifying as Digisexual. Here's What That Means." *Paper*, January 19, 2020. https://www.papermag.com/digisexual-sex-with-sandra-2644656349.html.

Strikwerda, Litska. 2017. "Legal and Moral Implications of Child Sex Robots." In *Robot Sex: Social and Ethical Implications*, edited by John Danaher and Neil McArthur, 133–151. Cambridge, MA: MIT Press.

Trayner, David. 2017. "First Sex Doll Brothel in Europe Shut Down One Month after Opening before Police Raid." *Daily Star*, March 17, 2017. https://www.dailystar.co.uk/news/latest-news/lumidolls-europe-first-sex-robot-16999176.

Turkle, Sherry. 1995. *Life on the Screen: Identity in the Age of the internet*. New York: Simon and Schuster.

———. 2007. "Authenticity in the Age of Digital Companions." *Interaction Studies* 8, no. 3: 501–517.

Wasserman, Eve. [Dr. Eve]. 2015. *Cyber Infidelity: The New Seduction*. Cape Town: Human and Rousseau.

Weichel, Andrew. 2020. "Sex Doll Rental Service Opening up in Metro Vancouver this Week." *CTV News*, January 6, 2020. https://bc.ctvnews.ca/mobile/sex-doll-rental-service-opening-up-in-metro-vancouver-this-week-1.4755977?fbclid=IwAR0nPkMgS5civ_IXB4JafCVjGSV0Ia9a_7F6eV6WQtVrfcLoWQBAPnZ2buA.

Wiseman, Eva. 2015. "Sex, Love and Robots: Is This the End of Intimacy?" *The Guardian*, December 13, 2015. https://www.theguardian.com/technology/2015/dec/13/sex-love-and-robots-the-end-of-intimacy.

Zimmer, Tyler. 2018. "Think Americans Should Have More Sex? Redistribute Wealth, Not Robots." *Slate*, May 4, 2018 https://slate.com/human-interest/2018/05/think-americans-should-have-more-sex-redistribute-wealth-not-robots.html.

Chapter 9

Robots as Caretakers

Understanding Long-Term Relationships between Humans and Carebots

Jamie Foster Campbell and Kristina M. Green

Advances in technology and science are increasing life expectancy globally, but younger generations are unable to meet the care needs of both aging populations and younger ones too. In Japan, for example, there is a social awareness that those populations who will require specialized care far outnumber the available caregivers who can undertake their diverse needs (Gallagher, Nåden, and Karterud 2016). In the United States, the Census Bureau projects that by 2050 "one in five Americans will be aged 65 or older, and at least 400,000 will be 100 or older" (as cited in Pew Research Center 2013, para. 1). Additionally, the "United Nations Population Division estimates for mid-2010, there were 642 million persons ages 0–4 and 523 million ages 65+" (Haub 2011, para. 2). In other words, and as of 2010, one in six people globally required specialized care, and considerations for how these responsibilities are met, and by whom, are of special interest in this chapter.

The purpose of this chapter is to offer a brief history of how robots became social, to discuss previous research on autonomous caretaking robots and to argue that scholars should reference interpersonal communication frameworks to better understand the possibility for long-term human–carebot relational development. We address issues of human vulnerability, what constitutes "good" care, and highlight the expansion of carebots in real-world settings, a transition that was accelerated by the COVID-19 global pandemic. The robotics industry may offer novel solutions to these problems through the integration of caretaking robots in society. As Broadbent, Stafford, and MacDonald (2009, 319) explain, "a healthcare robot is primarily intended to improve or protect the health and lifestyle of the human user." For example, carebots can deliver meals, administer medication, take vitals, do laundry, or offer companionship (Broadbent, Stafford, and MacDonald 2009; Kim, Park, and Sundar 2013). This work explores the paradoxical relationship between carebots as technologies that can extend life expectancy, independence, and remedy loneliness, while simultaneously ushering in an era of potential human insecurity and exploitation.

Crenshaw's (1989, 1990) framework on *intersectionality* offers a critical lens to examine concerns about vulnerability, care, power, and praxis as social robots are enlisted to help solve public health crises (see also Bowleg 2020, 2021; Collins 2015). In her canonical work about "the marginalization of Black women in feminist theory and antiracist politics," intersectionality refers to the limitations of a single-axis analysis that "does not accurately reflect the interaction of race and gender" (Crenshaw 1989, 140). Collins (2015, 1) adds that "the term intersectionality references the critical insight that race, class, gender, sexuality, ethnicity, nation, ability, and age operate not as unitary, mutually exclusive entities, but rather as reciprocally constructing phenomena." Furthermore, Crenshaw likens intersectionality to a prism: "a lens [. . .] for seeing the way in which various forms of inequality often operate together and exacerbate each other" (Steinmetz 2020, para. 3)—what Razack (1998, 14) refers to as "interlocking" relationships. In the context of social robots, there are many forms of interlocking inequality including race, "gender, class, sexuality, or immigrant status" (Steinmetz 2020, para. 3), as well as issues of digital disparities and barriers to access, which are discussed later in this chapter.

Carebots as an innovation for helping to solve problems related to the growing need for caregivers implicates discourses about what constitutes human vulnerability and care. Therefore, we begin with the premise that human vulnerability is contextual and historically informed (Malgieri and Niklas 2020). Writing about older populations, Gallagher, Nåden, and Karterud (2016) suggest that part of vulnerability includes reliance on others to respond to their physical, emotional, and cognitive needs. Additionally, the authors continue that older adults, for example, rely on caregivers "to have the ability to pick up on subtle cues regarding capabilities and to communicate in an emotionally engaged and meaningful way" (Gallagher, Nåden, and Karterud 2016, 369). These social needs found in elder care can also be spotted across other vulnerable populations and the care they require as well.

Borrowing from Luna's (2009) metaphor of *layers of vulnerability*, our identities are not fixed or static. Our vulnerability fluctuates and is "constructed by status, time and location. In this sense, the concept of layering provides an opening to a more intersectional approach and stresses its cumulative and transitory potential" (Malgieri and Niklas 2020, 4). Luna's (2009) conceptualization of layers of vulnerability reminds us that vulnerability is situated or nomadic—two key constructs explicated in feminist literature including *situated knowledge* (Haraway 1988; Suchman 2007), *nomadism* (Braidotti 2012), and *standpoint theory* (Harding 2004). To illustrate the idea of layers of vulnerability, Luna explains:

> if the situation of women is considered, it can be said that being a woman does not, in itself, imply that a person is vulnerable. A woman living in a country that does not recognize, or is intolerant of reproductive rights acquires a layer of vulnerability (that a woman living in other countries that respect such rights does not necessarily have). (2009, 128–29)

Additionally, it is not a matter of the presence or absence of vulnerability, but the degrees to which our vulnerability oscillates over time (e.g., aging), across geographical locations (e.g., governments and culture), and according to our perceived relative status to others.

Like vulnerability, care is also an overly broad term that encompasses multiple meanings. In her theory of post-human care, DeFalco (2020) interrogates the taken-for-granted assumptions about what constitutes good human care. "In the broadest sense," DeFalco (2020, 33) writes, "care is affection, devotion, responsibility, even obligation; it is action, behaviour, motivation and practice: care feels and care does." Care is at once familiar, abstract, and material and, as DeFalco suggests, highly personal. With these ideas in mind, carebots have been introduced as potential social actors for mitigating issues of caregiving capabilities in light of the sheer number of people who require care now and in the future. This crisis of care has captured the attention of roboticists, healthcare professionals, ethicists, and policymakers looking for practical solutions to a difficult and sensitive problem.

For long-term relationships to develop between humans and carebots, the field of communication can contribute important frameworks for rearticulating robots as social and communicative technologies that people collaborate with as opposed to utilities or tools that we use (Gunkel 2017; Guzman 2018; Šabanović and Chang 2016; Spence et al. 2014). Social robots are defined as embodied agents that can explicitly communicate with and learn from each other, as well as human users (Fong, Nourbakhsh, and Dautenhahn 2003). Breazeal (2003, 168) expands this definition and explains that social robots are "autonomous robots [who] perceive their world, make a decision on their own, and perform coordinated actions to carry out their task." Aymerich-Franch and Ferrer (2020) add that social robots imitate human behavioral norms and social cues during communicative exchanges. Caretaking robots or carebots, therefore, refers to social robots that are designed for the healthcare industry. Referring to robotics more broadly, Turkle (2011, 30) frames this "new breed of robots" as something to grow old with, machines that work alongside humans.

This chapter is organized as follows. First, we provide a brief background of human–machine communication (HMC). According to Guzman (2018, 11), "in HMC, technology is conceptualized as more than a channel or medium: it enters into the role of a communicator." Whereas the standard definition of communication refers to meaning made between people in and through their interactions with one another, HMC refers to "the creation of meaning among humans and machines. It is a process in which both humans and machines are involved and without one or the other communication would cease" (Guzman 2018, 17). Second, we highlight past research on social robots used for caretaking purposes. This foundation is necessary for scholars to understand how robots have been and will be used for long-term caregiving. Third, we use interpersonal communication literature to unpack what is needed for the development of long-term relationships between humans and carebots, specifically focusing on the evolution of trust and self-disclosure. Finally, we address ethical questions social scientists and roboticists should consider regarding the formation of care-based relationships between humans and robots over time.

BACKGROUND: HUMAN–MACHINE COMMUNICATION

The prospect of imbuing machines with human-like intelligence captured researchers' curiosity beginning in the 1950s with William Grey Walter's robotic tortoises and Alan Turing's acclaimed Turing Test. In the 1970s, Joseph Weizenbaum created a natural language processing program called ELIZA, one of the first conversational agents to imitate reciprocity by responding to questions posed by people. In other words, ELIZA simulated conversation through turn-taking. From early examples like ELIZA, machines were programmed to produce a desired response without understanding the meaning of what was being asked. These early examples of social robots were created with the goals of attempting to replicate human intelligence and increase cultural acceptance of intelligent machines. By 1950s standards, the benchmark of human intelligence was equated to animation; by the 1970s, it was equated to speech. Today, the benchmark has entered the realm of cognitive computing, including attempts to replicate human emotions and affect.

Beginning in the 1990s, Dr. Cynthia Breazeal created Kismet, a machine that learns and mimics human facial emotions. Through the movement of its mechanical eyes, ears, mouth, and so on, Kismet's emotive responses elicit strong reactions from the people who interact with it (Breazeal 2003, 2004). This is an example of a persuasive and affective machine. What is interesting about Kismet is its design and ability to learn from the feedback humans provided during their interactions and to imitate humans' expressiveness in real-time (Breazeal 2003). Contemporary carebots are replicating machine-learning techniques inherent to Kismet's design. In these ways, the interactions between humans and social robots are temporal: communication between humans and machines in the past and present dictate interactions, behaviors, and possibilities for the future. While machines today still don't understand the meaning or context of what is communicated, computation has come a long way in terms of robots that can imitate, calculate, infer, create, and recommend responses that were never programmed. Through advances in artificial intelligence (AI), social robots have grown in both sophistication and autonomy.

Robots like Boston Dynamics' ATLAS, designed for search and rescue, and Softbank Robotics' Pepper, a humanoid robot designed to read emotions, are driving the market toward even more autonomous social robots that can enhance people's lives, facilitate relationships, and potentially expand surveillance. The realities of today are that machines have been made artificially intelligent to cope with an overabundance of data, the scale and scope of which is so unwieldy that humans are unable to make sense of it manually (Kitchin 2014). Intelligent machines already represent a consequential and permanent feature of our everyday lives, but more recently, the persuasive and affective potentialities of social robots distinguish them as the next "new media" (Guzman 2018, 11). Therefore, we direct our attention to how interpersonal communication with these machines is meaningful (Guzman 2018).

Reeves and Nass (1996) laid the groundwork for understanding how people respond to computers as social actors (i.e., the CASA[1] paradigm). According to CASA, if the proper social cues are present during HMC, people respond to machines in the same ways they respond to people. In other words, people apply social rules and expectations to technology. Additionally, Turkle (2011, 11) argues that bonds with robots are possible as "we are psychologically programmed not only to nurture what we love, but to love what we nurture." For instance, Bickmore and Picard (2005) discovered that Bandai's Tamagotchi users reported having an emotional connection to their robotic pet and considered them part of their family. Similarly, 26 percent of users view Sony's AIBO as a companion and report experiencing an emotional connection to their robot dog, even missing AIBO when apart (Friedman, Kahn, and Hagman 2003). These examples illustrate that long-term relationships are possible with robotic entities.

Our relationship with machines becomes more complicated if we admit that machines have agency (Sandry 2015). Agency is an ambiguous concept that has taken on various meanings. Traditionally, this word refers to the human capacity to act freely. However, Suchman (2007) believes human- and machine-agencies coexist and are, therefore, not diametrically opposed.

Furthermore, if we put aside an anthropocentric perspective that only humans can be "intelligent," then multiple forms of agency that operate at once can enhance our understanding of HMC. Guzman (2018, 11) explains that in HMC "the machine is theorized as having a degree of agency in that it performs a distinct role during an interaction and draws from its own resources in processing and responding to messages." By rethinking the question of machine agency, we can better articulate how humans and machines co-construct the world we live in together (Guzman 2017; Sandry 2015).

Social Robots as Caregivers

Machine autonomy refers to the software and hardware capabilities that make machine learning, computer vision, and task completion possible without human intervention (Hakli and Mäkelä 2016). According to Hakli and Mäkelä (2016, 146), "what makes an agent autonomous is its capacity for perceiving and acting in its environment and learning from its experiences in order to modify its behaviour to promote its survival and improve its performance in achieving its tasks." Today, Japan leads in the production of carebots, because they are uniquely and contemporarily problem-solving the reality of their aging population. For example, Honda's ASIMO robot,[2] a partially autonomous humanoid robot, can assist with food preparation, recognize voices, gestures, and objects and can lift a patient in and out of bed. Researchers predict that carebots will be used for a wider range of caretaking tasks (e.g., assisting with baths, lifting heavy objects, and so on), as well as relational tasks (e.g., social companionship or recording personal histories; Bickmore and Picard 2005; van Wynsberghe 2013). Furthermore, Šabanović and Chang (2016) believe that reinforcing the social interaction between humans and machines will facilitate the acceptance of carebots among the average person (i.e., the more exposure people have to social robots, the more likely they will accept their help over time).

As carebots become increasingly part of our social world, people will base their interactions with them on existing interpersonal relationships and norms, as well as past experiences with and depictions of robots in popular culture. Individuals draw on these experiences to guide their understanding of how to act and communicate with machines (Guzman 2017). However, HMC is not universal. Bartneck and colleagues (2007), for example, discovered that different cultures have different levels of exposure to robots through media, and these disparate cognitive scripts influence humans' perceptions and future willingness to interact with them. Prior exposure to robots, whether through popular culture or first-hand experience, plays a significant role in determining our attitudes about social robots, and ultimately influences the relationships people form with these technologies as they become more immersed in our lives (Bartneck et al. 2007). The purpose and context of social robots (e.g., companionship, healthcare, sanitation, customer service, surveillance, etc.) dictate how humans will behave with and around these machines.

In light of the aforementioned history and operationalized terms, we now offer three carebot examples utilized in healthcare today: PARO, KASPAR, and RIBA. First, PARO, an interactive robot seal designed by the Japanese company AIST in 1993, represents a carebot designed to help relieve stress and improve socialization among elderly patients with dementia (Šabanović and Chang 2016; Severinson-Eklundh, Green, and Hüttenrauch 2003). Since becoming commercially available in the United States in 2005, researchers have employed this therapeutic robot as an object of study to learn more about its affordances and perceptions among patients, family, and staff alike. For instance, Šabanović and Chang (2016) discovered that social interactions with PARO in a large group setting (i.e., within the larger community of a nursing home) produced favorable attitudes and feelings of trust toward PARO.

Second, KASPAR (Kinesics and Synchronization in Personal Assistant Robotics) is a humanoid carebot designed to function as a social companion for children with autism. The University of Hertfordshire's Adaptive Systems Research Group started the KASPAR project in 2005 to create a carebot that acts as a mediator between children with autism and others (e.g., parents, teachers, classmates, therapists; Huijnen, Lexis, and de Witte 2016). Similar to PARO, KASPAR serves as a therapeutic and learning companion that encourages social development. Turkle (2005) asserts that there is value in research and implementations of therapy carebots. However, she warns that HMC cannot replace human relationships altogether nor should the technical, physical, and social capabilities of these machines dispossess us of our responsibilities to others (Turkle 2005).

Finally, RIBA (Robot for Interactive Body Assistance), developed by researchers at RIKEN and Tokai Rubber Industries in 2009, is a robotic nurse that uses tactile sensors to lift and carry patients and resembles a teddy bear to enhance perceptions of friendliness. Designed to serve as a healthcare assistant when a team of people is not available, RIBA is a descendant of industrial robotics and was created in response to the demand of Japan's aging population (Böhlen and Karppi 2017). Böhlen and Karppi (2017, 10) explain "RIBA is the first robot to demonstrate the ability to gracefully lift and move live human beings comfortably in and out of a bed in hospital practice." Together, these examples show how carebots are used for therapeutic and medical care with promising results.

While Japan pioneered carebots, the COVID-19 global pandemic ensured that they diffused globally, offering unique, complex, and transnational opportunities and consequences. According to their work on the implementation of carebots during this pandemic, Aymerich-Franch and Ferrer (2020) found that there were sixty-six different social robots deployed worldwide (e.g., Robotemi's Temi, Zorabots' James, and Softbank's Pepper). In their report, these social robots were used across thirty-five nations (e.g., China, United States, Thailand, Belgium, Hong Kong, India, and The Netherlands) and were implemented across a wide range of public sector institutions such as hospitals, nursing homes, schools, airports, hotels, office buildings, and more (Aymerich-Franch and Ferrer 2020).

Aymerich-Franch and Ferrer (2020) suggest that throughout the pandemic social robots serve a strategic role in not only facilitating physical distance between people but also functioning as (1) *liaisons* for human–human communication, (2) *safeguards* to reduce risks among healthcare workers, and (3) *well-being coaches* that offer support for quarantined or ill patients. For example, social robots that function as liaisons are described by the authors as delegating responsibilities traditionally held by receptionists in hospitals or customer service representatives at airports. During COVID-19, the health sector saw an increase in social robots handling pre-diagnosis or health monitoring tasks (e.g., questionnaires and thermal screenings) and facilitating telepresence between patients, medical personnel, and families since many of these technologies are equipped with cameras, displays, and audio capabilities (Aymerich-Franch and Ferrer 2020).[3]

In terms of safeguarding, the authors suggest that the rise of social robots reduces the risk of transmission, particularly among healthcare professionals that constituted upward of 25 percent of positive infection rates in the European Union and European Economic Area (ECDC 2020). Safeguarding tasks include disinfecting public spaces and communicating health safety advice to the public. Additionally, "[w]hen there is a safety concern regarding these aspects, the robot can take actions such as asking the person to wear a mask, keep physical distancing, or ban entrance to an indoor space" (Aymerich-Franch and Ferrer 2020, 6). Finally, Aymerich-Franch and Ferrer (2020) describe social robots as well-being coaches that offer companionship, entertainment, edutainment, and sometimes "nudging" or reminding people to exercise, and so on.

ROBOTIC COMPANIONS: TRUST, SELF-DISCLOSURE, AND SOCIAL SUPPORT

Interpersonal communication literature explains that trust and reciprocal self-disclosure are important for the early stages of relational development; they are considered the cornerstones for building companionships (Altman and Taylor 1973). Trust encompasses many dimensions; no one definition exists in the literature. However, the consensus among scholars suggests that trust is a dynamic, ongoing, and multidimensional construct (Barber 1983; Lewicki, Tomlinson, and Gillespie 2006). Simpson (2007) argues that interpersonal trust is anchored in personal goals and motives, which is why trust changes based on the context and relationship type. Furthermore, individuals have expectations, preferences, and needs, which all affect the development of trust. Trust is built over time and involves collaboration—for a relationship to survive, trust cannot be one-sided (Barber 1983).

Trust grows through self-disclosure: sharing personal information with others (Altman and Taylor 1973). Self-disclosure creates feelings of connection and facilitates bonding (Duck 2007; Stafford and Canary 1991). Within health care, self-disclosure from a care provider must maintain a balance between professionalism and openness (Dowling 2006). Similarly, trust between patients and caregivers is predicated on knowledge, skills, honesty, and competence (Rutherford 2014). In the caregiver–patient relationship, trust and self-disclosure increase if patients feel their views are respected and the information about their healthcare needs is transparently communicated (Rowe and Kellam 2010).

Previous research demonstrates that people feel comfortable self-disclosing personal information with robotic companions (Turkle et al. 2006). Even though people recognize that bonding with carebots is different from bonding with others, studies conclude that people feel connected with objects that make them feel cared for and accepted (Bickmore and Picard 2005; Turkle et al. 2006). Current research demonstrates that increased trust can lead to patients confiding in carebots, for example telling a carebot stories or discussing personal issues (Bickmore and Picard 2005; Martelaro et al. 2016).

For instance, Turkle and colleagues (2006) completed a study with *My Real Baby*, a robotic baby doll, and elderly patients in nursing homes. They discovered that residents reported feeling more comfortable confiding in *My Real Baby* (i.e., sharing personal problems) than with a person (Turkle et al. 2006). For example, seventy-four-year-old Jonathan explained that he would feel less embarrassed talking to a robot about something highly private, suggesting that the robot would not criticize him as a person might. Another participant, seventy-six-year-old Andy, explained feeling more comfortable expressing regret and frustration to a robot. Turkle and colleagues (2006, 357) emphasize how Andy "enjoys showing the doll to visitors and introduces her almost as one would a family member." In this interaction, the doll made Andy, who suffers from anxiety and depression, laugh and provided him with social support. Martelaro and colleagues (2016, 1) elaborate on this idea, suggesting that "eliciting self-disclosure may lead to stronger companionship between people and robots," as well as provide the social and emotional support needed to express levels of vulnerability that are inherent to the human condition.

Similarly, researchers found positive health outcomes for patients who interacted with carebots such as PARO. For example, patients with dementia who shared stories and discussed personal issues with PARO exhibited improved mental health, diminished loneliness, and perceptions of social support (Šabanović and Chang 2016; Severinson-Eklundh, Green, and Hüttenrauch 2003; Turkle 2005). In this way, carebots may be perceived as "reliable companions" (Kim, Park, and Sundar 2013, 1799).

Interpersonal communication scholarship can help create a foundation to conceptualize how people and carebots can establish long-term relationships. When considering interpersonal relationships, the reciprocal nature of trust and self-disclosure are important. With the aforementioned interpersonal communication constructs in mind, the following questions arise: How is trust conceptualized in human–carebot relationships? In a healthcare setting, will humans perceive carebots as reciprocating a form of machine-self-disclosure? Does the illusion of reciprocal self-disclosure and trust performed by machines threaten people's perceptions of these key relational characteristics? And if so, do people's renewed or diminished interpretation of relational trust transfer or map onto their human–human relationships thereafter?

With this in mind, scholars are urged to consider what simulated self-disclosure between humans and carebots might entail. What is the consequence of programming carebots with personalities (see Samsung Newsroom US 2019)? Should these AI personalities change based on the personalities of their interactional partners? Do carebots need to become conversational agents (Gunkel 2017; Turkle 2011)? How can the information that carebots receive from patients be used to responsibly inform healthcare decisions or diagnoses? How will carebots manage patients who are stubborn, confused, aggressive, and violent?

So far the patient–carebot relationship has been the focal point for discussing constructs like trust and self-disclosure. Nevertheless, how carebots interact with a patient's extended network of family and friends is equally important. A patient's extended network of support plays a vital role in key decision-making, and they will also have interpersonal contact with carebots. While this topic warrants its own paper, more scholarship is needed to consider deeply how formal and informal caregivers (e.g., medical staff, family members, spouses, etc.) will also interact with the carebot (Draper and Sorell 2017). Previous research concludes that in the context of health care, the provider's demeanor affects patients and their families' perception of the care received (Ward-Griffin and McKeever 2000). With that said, new questions arise: To what extent will people trust and entrust their loved ones to carebots? What warrants appropriate and inappropriate disclosures of sensitive information captured by a carebot? What and to whom should a carebot report?

The more nuanced and less straightforward problem-solving aspects of HMC are the more interpersonal, cross-cultural, and ethical they become. "Socially interactive robots," according to Fong, Nourbakhsh, and Dautenhahn (2013, 148), "must also address those issues imposed by social interaction" (see also Breazeal 2004). The work being done today to design the next generations of carebots is heavily invested in engineering innovations that can sufficiently handle the spontaneity of interpersonal communication inherent to HMC. Time, effort, and resources should be dedicated toward asking and answering hard questions that reimagine the meaning of relational development, trust, and self-disclosure as machines begin occupying increasingly autonomous roles as caregivers, especially to vulnerable people.

A NEW ETHICAL LANDSCAPE

Caregiving is no longer a distinctively human act. The adoption of carebots in medical and therapeutic contexts is creating a new type of caregiving relationship. Contemporarily, we find ourselves at a moral and legal crossroads that demands preparedness for the challenges, risks, and social realities that people now coexist alongside and are growing to rely upon artificially intelligent carebots. To begin, the deployment of social robots, particularly in healthcare contexts, is quickly outpacing policy and legal frameworks that are already in place to protect patients. For instance, Annaswamy, Verduzco-Gutierrez, and Frieden (2020) conclude that, in the U.S. context, existing language from The Americans with Disability Act of 1990 does not extend to virtual spaces and demands careful amendment. Additionally, many accessibility laws intended to increase equity are voluntary. Scholars in the field of health policy and technology who research digital disparities recommend that the "[t]echnology companies that design and distribute telemedicine products must be subject to these laws, by considering them and their products as healthcare—not technological—organizations and products" (Annaswamy, Verduzco-Gutierrez, and Frieden 2020, 2). Similarly, and within an American context, the authors emphasize that without nation-wide telehealth regulatory frameworks, the existing "hodgepodge of conflicting local, state, insurance, and federal regulations" will disadvantage patients and healthcare workers alike (Annaswamy, Verduzco-Gutierrez, and Frieden 2020, 2).

In addition to the aforementioned legislative challenges, there is also the matter of digital and health disparities. Digital disparities refer to barriers of infrastructure (e.g., access to broadband Internet) that affect underserved and historically oppressed populations. According to their research on COVID-19 and digital inequity, Ortega et al. (2020, 368) found that those who face digital disparities include "racial/ethnic minorities, patients who live in rural areas, patients with limited English proficiency (LEP), with low literacy, or low income, [and where] telemedicine is often less accessible." Many countries across the world face digital disparity challenges in light of the COVID-19 pandemic.

To revisit Crenshaw's (1989, 1990) and Collins's (2015) work on inter-sectionality, related and pressing issues that demand greater attention are racialized health disparities. According to Halfon and colleagues (2020, 1702), and within the context of the United States, health disparities are "sig-nificant, persistent, and highly racialized health inequities rooted in complex historical, social, and structural factors—particularly at the nexus of race and income" that "begin early in life." In other words, research in medicine and public health have identified *racism*—not race—as a "key driver of the social, political, and economic injustices that cause and maintain health ineq-uities" (Truong and Sharif 2021, 1). Annaswamy, Verduzco-Gutierrez, and Frieden (2020) add that the widespread use of telehealth appointments during COVID-19 revealed widespread inaccessibility surrounding user interface design for persons with disabilities (e.g., screen reader, sign language, cap-tions, magnification, color, and contrast). The authors conclude that "[i]n the era of Covid-19 and beyond, telemedicine can no longer be considered a 'complement' to in-person care" (Annaswamy, Verduzco-Gutierrez, and Frieden 2020, 3). Lessons from telemedicine can be extended to carebots insofar that they both mediate in-person care when human-to-human contact is not safe or when resources for care are limited. In tackling these challenges and cultivating greater preparedness, a collective commitment to addressing these disparities ethically, especially in cases where vulnerable populations have their healthcare needs met by machines, is imperative.

Ethical Implications

The literature on robotic ethics and vulnerable populations has identified both advantages and disadvantages of HRI that are nuanced and context-specific (Beer et al. 2012; Feil-Seifer and Mataric 2011; Sharkey and Sharkey 2011; Sparrow and Sparrow 2006). For example, Beer and colleagues (2012) found social acceptance of robots among older adults when these technologies functioned as domestic helpers tasked with completing household chores. However, in the same study, older adults were less accepting of robots when they were tasked with monitoring or intended to function as companions. In this example, the tasks and responsibilities undertaken by the social robots co-construct the identity and positionality of the individual. A carebot that assists with household chores may be perceived as subordinate to the human being serviced by this behavior. By contrast, a carebot tasked with overseeing or accompanying an individual may be perceived as imposing or infringing on that person's sense of privacy, agency, or independence.

O'Brolcháin (2019) also considers the benefits and moral hazards of the adoption of carebots for people with dementia and society as a whole. On the one hand, carebots can help complete household chores like changing bed linens or help maintain routines like personal hygiene, reduce instances of abuse, and offer potential psychological benefits such as companionship (O'Brolcháin 2019). On the other, the author also foresees that people with dementia (PwD) may become even more isolated than they already are as people transfer the difficult responsibilities of care to machines. In this way, the diffusion of carebots may create a *moral hazard*: "a situation in which being able to act without bearing the costs of the consequences relieves agents of responsibility for their decisions" (Chamayou 2015; as quoted in O'Brolcháin 2019, 966). In this case, O'Brolcháin (2019, 966) raises the possibility of communities to "outsource care for PwD to non-human actors [making it] easier for societies using robots to abandon care of vulnerable people, such as PwD."

Finally, Sharkey and Sharkey (2011, 35) also identify the pros and cons of HRI that depend on variables like age because "both the young and the old show a strong tendency to anthropomorphize robot companions and pets." For children, the authors maintain an unlikelihood of positive outcomes resulting from interactions between infants and interactive robots. After considering the role that people play in helping children develop social attachment and emotional intelligence, the authors conclude that "spending too much time in the company of a robot is unlikely to help" facilitate empathy and emotional development (Sharkey and Sharkey 2011, 36). Additionally, the authors suggest that the insertion of social robots into this dynamic "could interfere with an infant's learning about the give and take of human relationships" (Sharkey and Sharkey 2011, 36–37).

For the elderly, Sharkey and Sharkey (2011) suggest HRI among Alzheimer's patients could go either way. For example, interactions with robotic pets could be reasonably said to facilitate the same kinds of positive health and emotional outcomes seen with doll therapy. Conversely, a relationship with a robot might increase anxiety if the patient feels responsible for their companion pet's well-being. In the latter case, the authors highlight an ethical dilemma: "observers and relatives of a confused [elderly] person looking after a robot pet might see it as depriving their relative of dignity and infantilizing them" (Sharkey and Sharkey 2011, 35). Thus, despite some potential benefits, the authors go on to identify six main ethical concerns within the context of HRI and elderly care: (1) reduced human contact, (2) increased objectification and loss of control, (3) loss of privacy, (4) loss of personal liberty, (5) deception and infantilization, and (6) questions of responsibility and circumstances when the elderly should be allowed to control carebots (Sharkey and Sharkey 2011).

In the aforementioned examples, an enormous amount of faith and responsibility has been placed on designers to understand and foresee the needs and challenges of vulnerable people who come into contact with carebots in medical and therapeutic settings (O'Brolcháin 2019). The meaning made between humans and machines during these interactions is not only foundational to HMC scholarship but raises important ethical questions about the kind of norms and values carebots can be made to support or undermine. Will elderly populations gain or lose their sense of independence with the help of carebots? Will vulnerable populations experience greater isolation if carebots become commonplace? Will society be less willing to care for elderly or sick people if they can transfer caregiving responsibilities to machines? Will the absence of these populations in society change our understanding of aging and death? Will the diffusion of carebots change the demands and cost of labor in nursing or related healthcare fields? How can carebot designers tackle bias? Questions like these highlight the significance of identifying values that can be integrated into the design of ethical carebots.

Designing Ethical Carebots

Ethical approaches to carebot design are primarily concerned with how to develop innovations that avoid harm and instead uphold the dignity and well-being of people. Draper and Sorell (2017) identify six key ethical values to incorporate into the design, development, and integration of robots tasked with caring for elderly people in their homes. These values— "autonomy, safety, enablement, independence, privacy, and social connectedness"—represent key criteria that roboticists should keep in mind and strive to enhance for others (see Draper and Sorell 2017, 49).

In the context of COVID-19, Vitak and Zimmer (2020) question how to best balance contextual risks and the need for short- and long-term privacy protections. Discussing contact tracing applications, the authors suggest that the most important privacy concern, "is the appropriateness of data flows from users' smartphones: who can access data, how long is data stored, and for what purposes could that data be used in the future" (Vitak and Zimmer 2020, 2). This implies that major concerns involve the risk between tracking location-based and health-related data alongside fears that once the pandemic is over, new surveillance norms will be left in its wake. These strategies for obtaining more information during times of crisis are not unique to the COVID-19 pandemic. What becomes a matter of significance is how this global health crisis normalizes and increases state surveillance (Eck and Hatz 2020, 609).

According to a 2020 United Nations (UN) report, the potential for (mis) information control and abuses of power are high, and "what is justified during an emergency now may become normalized once the crisis has passed. . . . All measures must incorporate meaningful data protection safeguards, be lawful, necessary, and proportionate, time-bound and justified by legitimate public health objectives" (UN 2020, 16). Scholars in media studies refer to this phenomenon as a type of digital *mission creep* (Eck and Hatz 2020): the tendency for government agencies and others to strive for *total information capture* (Andrejevic 2019) as opposed to restricting massive data collection.

These patterns of surveillance, digital creep, and privacy protection concerns amplified by COVID-19 cross-pollinate to social robots and the crisis of care that societies around the globe are contending with. While Vitak and Zimmer (2020, 3) acknowledge that companies like Google and Apple are working toward better data privacy governance, the cultural acquiescence to widespread surveillance measures as a result of COVID-19 (e.g., thermal cameras and Bluetooth monitoring) may "be used to justify wider escalations of health monitoring that impinge on individual autonomy." Since carebots capture intimate data about patients' biometrics, diagnoses, treatments, and protected health information, these concerns over health monitoring and autonomy extend to research and design for ethical carebots.

With carebots, we see that the core ethical considerations are familiar constructs being asked in new contexts. Ess (2017, 9) explains, as carebots are designed to "replicate embodied human beings in a number of ways, robots further implicate an entire range of philosophical questions, beginning with our understandings of human identity and agency, the role of emotions in communication and ethical decision-making, and so on." As carebots diffuse, we also need to consider their limits. For example, how should a carebot behave when a terminally ill person wishes to die? Böhlen and Karppi (2017, 3) ask, "what kind of last comforts and last wishes should a system deliver, and what should it never do, despite its abilities?" Additionally, van Wynsberghe (2013) argues that although ethical systems need to be built into carebots, humans will also need to consider ethical ways to interact with their carebot. Should carebots be held responsible for the decisions they make or the actions they initiate? If not, who undertakes this responsibility, both legally and morally?

Some researchers argue that we have surpassed traditional frameworks for civil and criminal negligence and even the introduction to new categories of personhood (e.g., "electronic personalities"; see Beck 2016). While these topics are outside the scope of this chapter, carebots are among a new constellation of technologies for which there are few legal and policy precedents. From a caregiving context, future scholars will need to consider healthcare regulations. What new privacy and electronic security measures need to be taken? If confidentiality is compromised, who should be held accountable? How would existing laws about power of attorney operate within a carebot context? In the United States, how might carebots be designed for HIPPA (Health Insurance Portability and Accountability Act) compliance? At the transnational level, how can carebots be designed to ensure that they meet the legal, ethical, and privacy standards dictated by their country or bloc?

CAREBOTS: WHERE WE GO FROM HERE

There is a lack of empirical research on the long-term relationships between patients and carebots, especially when involving multiple stakeholders (e.g., healthcare professionals, patients, policymakers, caregivers, and citizens; Bickmore and Picard 2005; Gallagher et al. 2016). From an HMC perspective, future researchers should reexamine their existing assumptions about the role of machine intelligence operating in healthcare settings. We argue that social robots are more than tools or channels; they are nonhuman communicators and consequential decision-makers, which raises questions about what it means to be *human* (see Guzman 2018) and what it means to be vulnerable. From trust and self-disclosure to remedying loneliness and challenging assumptions about agency, carebots represent a new communicative partner that reimagines what caregiving entails and who is responsible for performing care-related tasks.

In this chapter, we discussed the global crisis of care and social robots as one practical but risky solution to a multifaceted and urgent social problem. During global health crises like Ebola and COVID-19, social robots have been deployed into real-world contexts as a necessary means for reducing risk, while also extending a new genre of care to others through machines and at a distance. A review of historical and contemporary literature suggests that the benchmarks for machine intelligence have shifted as AI technology becomes more ubiquitous and advanced. Whereas machine intelligence denoted animation and speech as key indicators of human intelligence, today's most advanced carebots are leveraging cognitive computing techniques that aim at replicating and/or facilitating human affect, emotion, and touch.

Traditionally, core needs concerning the physical, emotional, and cognitive well-being of humans were attended to by other people. Contemporarily, however, social robots have entered the fold and increased the likelihood that our needs may be met by nonhuman others. In this way, social robots are both descended and distinct from past telemedical approaches. *Tele-* from the Greek "far," denotes that telemedicine has always implicated care at a distance. In the past, this included telephone appointments between patients and doctors or televised public health campaigns. The distinguishing factor from these interpersonal and mass communication examples is that artificially intelligent machines represent persuasive and personalized media that can use health information to track, learn, predict, and recommend courses of action for the future based on the past. Today, social robots do not merely exchange information but create and co-construct data.

In light of the larger questions posed throughout this chapter, a reimagining of how we think about relational development between people and carebots is needed. We cannot expect that reciprocal self-disclosure, or the development of trust, will look the same in HMC as it does in our interpersonal relationships. As people become more dependent on carebots, interdisciplinary theories and methods (e.g., communication, Black feminist epistemology, computer science, ethics, law and justice, medicine, psychology, etc.) become important for developing more equitable technology in the healthcare industry (Feil-Seifer and Mataric 2011).

As Bowleg (2021, 89) suggests, an intersectional approach to public health identifies research as "an important step on the journey to health equity [. . .] not the destination." Research can help identify moral hazards and steer us toward potential solutions but an intersectional approach to carebots demands that scholars put the cumulative knowledge found across disciplines into action. Therefore, a multi-axis approach to ethical design—one that considers digital and health disparities—must contend with the reality that technology is not neutral and can disenfranchise or threaten the dignity and livelihoods of vulnerable persons.

As humans increasingly socialize with machines, borrowing interdisciplinary theories and methodologies may help inform decision-making and ethical design. The application of carebots will reinvent not only institutions of health and medicine but social norms and expectations about privacy and surveillance too. We are at the dawn of a new era and the aforementioned queries are worthy of further exploration. While there are more questions than answers, this topic requires all of us, scholars and non-, to think critically about why, how, and under what circumstances carebots can (or should) replace people and the extent to which social robots can be made to foster long-term caregiving relationships.

NOTES

1. Computers Are Social Actors — Ed.
2. Honda's ASIMO robot was discontinued in 2018, but its inclusion here is still relevant and important as it represents how state of the art these technologies are today. From the time this manuscript was conceived, drafted, and finally published the carebots under discussion will inevitably change in significant ways.

3. Hanson Robotics' Sophia, possible the most famous—and fraught—social robot at the time of writing (see also Empey, chapter 6 and Ley, chapter 7 this book), is also being released commercially in 2021 "for use in healthcare, retail, and airline settings" (Pauly 2021). Sophia and siblings are being framed by Hanson as "so useful during these times where people are terribly lonely and socially isolated" (quoted in Pauly 2021). — Ed.

REFERENCES

Altman, Irwin, and Dalmas A. Taylor. 1973. *Social Penetration: The Development of Interpersonal Relationships*. Social Penetration: The Development of Interpersonal Relationships. Oxford, England: Holt, Rinehart & Winston.

Andrejevic, Mark. 2019. "Automating Surveillance." *Surveillance & Society* 17 (1/2): 7–13. https://doi.org/10.24908/ss.v17i1/2.12930.

Annaswamy, Thiru M., Monica Verduzco-Gutierrez, and Lex Frieden. 2020. "Telemedicine Barriers and Challenges for Persons with Disabilities: COVID-19 and Beyond." *Disability and Health Journal* 13 (4): 100973. https://doi.org/10 .1016/j.dhjo.2020.100973.

Aymerich-Franch, Laura, and Iliana Ferrer. 2020. "The Implementation of Social Robots during the COVID-19 Pandemic." University of Cornell Preprint. http://arxiv.org/abs/2007.03941.

Barber, Bernard. 1983. *The Logic and Limits of Trust*. New Brunswick: Rutgers University Press.

Bartneck, Christoph, Tomohiro Suzuki, Takayuki Kanda, and Tatsuya Nomura. 2007. "The Influence of People's Culture and Prior Experiences with Aibo on Their Attitude towards Robots." *AI & Society* 21 (1): 217–30. https://doi.org/10.1007 /s00146-006-0052-7.

Beck, Susanne. 2016. "Intelligent Agents and Criminal Law—Negligence, Diffusion of Liability and Electronic Personhood." *Robotics and Autonomous Systems* 86 (December): 138–43. https://doi.org/10.1016/j.robot.2016.08.028.

Beer, Jenay M., Cory-Ann Smarr, Tiffany L. Chen, Akanksha Prakash, Tracy L. Mitzner, Charles C. Kemp, and Wendy A. Rogers. 2012. "The Domesticated Robot: Design Guidelines for Assisting Older Adults to Age in Place." In *Proceedings of the Seventh Annual ACM/IEEE International Conference on Human-Robot Interaction*, 335–42. HRI '12. Boston, MA: Association for Computing Machinery. https://doi.org/10.1145/2157689.2157806.

Bickmore, Timothy W., and Rosalind W. Picard. 2005. "Establishing and Maintaining Long-Term Human-Computer Relationships." *ACM Transactions on Computer-Human Interaction* 12 (2): 293–327. https://doi.org/10.1145/1067860.1067867.

Böhlen, Marc, and Tero Karppi. 2017. "The Making of Robot Care." *Transformations* 29: 1–22.

Bowleg, Lisa. 2020. "We're Not All in This Together: On COVID-19, Intersectionality, and Structural Inequality." *American Journal of Public Health* 110 (7): 917–17. https://doi.org/10.2105/AJPH.2020.305766.

———. 2021. "Evolving Intersectionality within Public Health: From Analysis to Action." *American Journal of Public Health* 111 (1): 88–90. https://doi.org/10.2105/AJPH.2020.306031.

Braidotti, Rosi. 2012. *Nomadic Theory: The Portable Rosi Braidotti*. New York: Columbia University Press.

Breazeal, Cynthia. 2003. "Toward Sociable Robots." *Robotics and Autonomous Systems, Socially Interactive Robots* 42 (3): 167–75. https://doi.org/10.1016/S0921-8890(02)00373-1.

Breazeal, Cynthia L. 2004. *Designing Sociable Robots*. Cambridge: MIT Press.

Broadbent, Elizabeth, Rebecca Stafford, and Bruce MacDonald. 2009. "Acceptance of Healthcare Robots for the Older Population: Review and Future Directions." *International Journal of Social Robotics* 1 (4): 319. https://doi.org/10.1007/s12369-009-0030-6

Chamayou, Grégoire. 2015. *A Theory of the Drone*. Translated by Janet Lloyd. New York: The New Press House.

Crenshaw, Kimberlé. 1989. "Demarginalizing the Intersection of Race and Sex: A Black Feminist Critique of Antidiscrimination Doctrine, Feminist Theory and Antiracist Politics." *University of Chicago Legal Forum*, 139–67.

———. 1990. "Mapping the Margins: Intersectionality, Identity Politics, and Violence against Women of Color." *Stanford Law Review* 43 (6): 1241–99. https://doi.org/10.2307/1229039.

Collins, Patricia Hill. 2015. "Intersectionality's Definitional Dilemmas." *Annual Review of Sociology* 41 (1): 1–20. https://doi.org/10.1146/annurev-soc-073014-112142.

Dowling, Maura. 2006. "The Sociology of Intimacy in the Nurse-Patient Relationship." *Nursing Standard* 20 (23): 48–54. https://doi.org/10.7748/ns2006.02.20.23.48.c4070.

Draper, Heather, and Tom Sorell. 2017. "Ethical Values and Social Care Robots for Older People: An International Qualitative Study." *Ethics and Information Technology* 19 (1): 49–68. https://doi.org/10.1007/s10676-016-9413-1.

Duck, Steve. 2007. *Human Relationships*. London: SAGE.

Ess, Charles Melvin. 2017. "Digital Media Ethics." *Oxford Research Encyclopedia of Communication*. https://www.duo.uio.no/handle/10852/65162.

Eck, Kristine, and Sophia Hatz. 2020. "State Surveillance and the COVID-19 Crisis." *Journal of Human Rights* 19 (5): 603–12. https://doi.org/10.1080/14754835.2020.1816163.

European Centre for Disease Prevention and Control. 2020. "Rapid Risk Assessment: Coronavirus Disease 2019 (COVID-19) Pandemic: Increased Transmission in the EU/EEA and the UK—Eighth Update." https://www.ecdc.europa.eu/en/publications-data/rapid-risk-assessment-coronavirus-disease-2019-covid-19-pandemic-eighth-update.

Feil-Seifer, David, and Maja J. Matarić. 2011. "Socially Assistive Robotics." *IEEE Robotics Automation Magazine* 18 (1): 24–31. https://doi.org/10.1109/MRA.2010.940150

Fong, Terrence, Illah Nourbakhsh, and Kerstin Dautenhahn. 2003. "A Survey of Socially Interactive Robots." *Robotics and Autonomous Systems, Socially Interactive Robots* 42 (3): 143–66. https://doi.org/10.1016/S0921-8890(02)00372-X.

Friedman, Batya, Peter H. Kahn, and Jennifer Hagman. 2003. "Hardware Companions? —What Online AIBO Discussion Forums Reveal about the Human-Robotic Relationship." In *Human Factors in Computing Systems*, 273–80. ACM.

Gallagher, Ann, Dagfinn Nåden, and Dag Karterud. 2016. "Robots in Elder Care: Some Ethical Questions." *Nursing Ethics* 23 (4): 369–71. https://doi.org/10.1177/0969733016647297.

Grey Walter, William. 1950. "An Electro-Mechanical Animal." *Dialectica* 4 (3): 206–13.

Gunkel, David J. 2017. "The Other Question: Socialbots and the Question of Ethics." In *Socialbots and Their Friends: Digital Media and the Automation of Sociality*, edited by Robert W. Gehl and Maria Bakardjieva. New York: Routledge.

Guzman, Andrea L. 2017. "Making AI Safe for Humans: A Conversation with Siri." In *Socialbots and Their Friends: Digital Media and the Automation of Sociality*, edited by Robert W. Gehl and Maria Bakardjieva, 69–85. New York: Routledge.

———. 2018. "Introduction: What Is Human-Machine Communication, Anyway." In *Human-Machine Communication: Rethinking Communication, Technology, and Ourselves*, edited by Andrea L. Guzman, 1–28. New York: Peter Lang Publishing, Inc.

Hakli, Raul, and Pekka Mäkela. 2016. "Robots, Autonomy, and Responsibility." In *What Social Robots Can and Should Do*, edited by Johanna Seibt, Marco Nørskov, and Søren Schack Andersen, 145–54. Amsterdam: IOS Press.

Halfon, Neal, Efren Aguilar, Lisa Stanley, Emily Hotez, Eryn Block, and Magdalena Janus. 2020. "Measuring Equity from the Start: Disparities in the Health Development of US Kindergartners." *Health Affairs* 39 (10): 1702–9. https://doi.org/10.1377/hlthaff.2020.00920.

Haub, Carl. 2011. "World Population Aging: Clocks Illustrate Growth in Population Under Age 5 and Over Age 65." Population Reference Bureau. https://www.prb.org/agingpopulationclocks/.

Haraway, Donna. 1988. "Situated Knowledges: The Science Question in Feminism and the Privilege of Partial Perspective." *Feminist Studies* 14 (3): 575–99. https://doi.org/10.2307/3178066.

Harding, Sandra G. 2004. *The Feminist Standpoint Theory Reader: Intellectual and Political Controversies*. East Sussex: Psychology Press.

Huijnen, Claire A. G. J., Monique A. S. Lexis, and Luc P. de Witte. 2016. "Matching Robot KASPAR to Autism Spectrum Disorder (ASD) Therapy and Educational Goals." *International Journal of Social Robotics* 8 (4): 445–55. https://doi.org/10.1007/s12369-016-0369-4.

Kim, Ki Joon, Eunil Park, and S. Shyam Sundar. 2013. "Caregiving Role in Human—Robot Interaction: A Study of the Mediating Effects of Perceived Benefit and Social Presence." *Computers in Human Behavior* 29 (4): 1799–806. https://doi.org/10.1016/j.chb.2013.02.009.

Kitchin, Rob. 2014. "Big Data, New Epistemologies and Paradigm Shifts." *Big Data & Society*. https://doi.org/10.1177/2053951714528481.

Lewicki, Roy J., Edward C. Tomlinson, and Nicole Gillespie. 2006. "Models of Interpersonal Trust Development: Theoretical Approaches, Empirical Evidence, and Future Directions." *Journal of Management* 32 (6): 991–1022. https://doi.org/10.1177/0149206306294405

Luna, Florencia. 2009. "Elucidating the Concept of Vulnerability: Layers Not Labels." *IJFAB: International Journal of Feminist Approaches to Bioethics* 2 (1): 121–39. https://doi.org/10.3138/ijfab.2.1.121.

Malgieri, Gianclaudio, and Jędrzej Niklas. 2020. "Vulnerable Data Subjects." *Computer Law & Security Review* 37 (July): 105415. https://doi.org/10.1016/j.clsr.2020.105415

Martelaro, Nikolas, Victoria C. Nneji, Wendy Ju, and Pamela Hinds. 2016. "Tell Me More Designing HRI to Encourage More Trust, Disclosure, and Companionship." In *2016 11th ACM/IEEE International Conference on Human-Robot Interaction (HRI)*, 181–88. https://doi.org/10.1109/HRI.2016.7451750.

O'Brolcháin, Fiachra. 2019. "Robots and People with Dementia: Unintended Consequences and Moral Hazard." *Nursing Ethics* 26 (4): 962–72. https://doi.org/10.1177/0969733017742960.

Ortega, Gezzer, Jorge A. Rodriguez, Lydia R. Maurer, Emily E. Witt, Numa Perez, Amanda Reich, and David W. Bates. 2020. "Telemedicine, COVID-19, and Disparities: Policy Implications." *Health Policy and Technology* 9 (3): 368–71. https://doi.org/10.1016/j.hlpt.2020.08.001.

Pauly, Alexandra. 2021. "Sophia the Robot Could be Coming to a Store Near You." *Hyperbae*, January 27, 2021. https://hypebae.com/2021/1/sophia-humanoid-robot-hanson-robotics-mass-production-healthcare-coronavirus-covid-19-announcement.

Pew Research Center. 2013. "Living to 120 and Beyond: Americans' Views on Aging, Medical Advances and Radical Life Extension." Polling and Analysis. Washington, DC. https://www.pewforum.org/2013/08/06/living-to-120-and-beyond-americans-views-on-aging-medical-advances-and-radical-life-extension/.

Razack, Sherene H. 1998. *Looking White People in the Eye: Gender, Race, and Culture in Courtrooms and Classrooms*. Toronto: University of Toronto Press.

Reeves, Byron, and Clifford Ivar Nass. 1996. *The Media Equation: How People Treat Computers, Television, and New Media like Real People and Places*. New York: Cambridge University Press.

Rowe, Jimmy, and Charles Kellam. 2010. "Trust: A Continuing Imperative." *Home Health Care Management & Practice* 22 (6): 417–23. https://doi.org/10.1177/1084822309341255.

Rutherford, Marcella M. 2014. "The Value of Trust to Nursing." *Nursing Economics* 32 (6): 283–327.

Šabanović, Selma, and Wan-Ling Chang. 2016. "Socializing Robots: Constructing Robotic Sociality in the Design and Use of the Assistive Robot PARO." *AI & Society* 31 (4): 537–51. https://doi.org/10.1007/s00146-015-0636-1.

Samsung Newsroom US. 2019. "Personalizing Elder Care with Intuition Robotics and 'ElliQ.'" https://news.samsung.com/us/intuition-robotics-and-personalizing-elder-care/.

Sandry, Eleanor. 2015. *Robots and Communication*. New York: Palgrave Macmillan. https://doi.org/10.1057/9781137468376.0001.

Severinson-Eklundh, Kerstin, Anders Green, and Helge Hüttenrauch. 2003. "Social and Collaborative Aspects of Interaction with a Service Robot." *Robotics and Autonomous Systems, Socially Interactive Robots* 42 (3): 223–34. https://doi.org /10.1016/S0921-8890(02)00377-9.

Sharkey, Amanda, and Noel Sharkey. 2011. "Children, the Elderly, and Interactive Robots." *IEEE Robotics Automation Magazine* 18 (1): 32–38. https://doi.org/10 .1109/MRA.2010.940151.

Simpson, Jeffry A. 2007. "Foundations of Interpersonal Trust." In *Social Psychology: Handbook of Basic Principles*, edited by Arie W. Kruglanski and E. Tory Higgins, 2nd ed., 587–607. New York: The Guilford Press.

Sparrow, Robert, and Linda Sparrow. 2006. "In the Hands of Machines? The Future of Aged Care." *Minds and Machines* 16 (2): 141–61. https://doi.org/10.1007 /s11023-006-9030-6.

Spence, Patric R., David Westerman, Chad Edwards, and Autumn Edwards (2014). "Welcoming Our Robot Overlords: Initial Expectations About Interaction with a Robot." *Communication Research Reports* 31 (3): 272–80. https://doi.org/10.1080 /08824096.2014.924337.

Stafford, Laura, and Daniel J. Canary. 1991. "Maintenance Strategies and Romantic Relationship Type, Gender and Relational Characteristics." *Journal of Social and Personal Relationships* 8 (2): 217–42. https://doi.org/10.1177 /0265407591082004.

Steinmetz, Katy. 2020. "She Coined the Term 'Intersectionality' Over 30 Years Ago. Here's What It Means to Her Today." *Time*, February 20, 2020. https://time.com /5786710/kimberle-crenshaw-intersectionality/.

Truong, Mandy, and Mienah Z. Sharif. 2021. "We're in This Together: A Reflection on How Bioethics and Public Health Can Collectively Advance Scientific Efforts Towards Addressing Racism." *Journal of Bioethical Inquiry,* January 7, 2021. 1–4. https://doi.org/10.1007/s11673-020-10069-w.

Turing, Alan M. 1950. "Computing Machinery and Intelligence." *Mind: A Quarterly Review* LIX (236): 433–60. https://doi.org/10.1093/mind/LIX.236.433.

Turkle, Sherry. 2005. *The Second Self: Computers and the Human Spirit.* Cambridge: MIT Press.

———. 2011. *Alone Together: Why We Expect More from Technology and Less from Each Other.* New York: Perseus Books Group.

Turkle, Sherry, Will Taggart, Cory D. Kidd, and Olivia Dasté. 2006. "Relational Artifacts with Children and Elders: The Complexities of Cybercompanionship." *Connection Science* 18 (4): 347–61. https://doi.org/10.1080/09540090600868912.

United Nations. 2020. "COVID-19 and Human Rights: We Are All in This Together." United Nations. https://www.un.org/en/un-coronavirus-communications-team/ we-are-all-together-human-rights-and-covid-19-response-and.

van Wynsberghe, Aimee. 2013. "Designing Robots for Care: Care Centered Value-Sensitive Design." *Science and Engineering Ethics* 19 (2): 407–33. https://doi.org /10.1007/s11948-011-9343-6.

Vitak, Jessica, and Michael Zimmer. 2020. "More Than Just Privacy: Using Contextual Integrity to Evaluate the Long-Term Risks from COVID-19 Surveillance

Technologies." *Social Media + Society* 6 (3): 2056305120948250. https://doi.org/10.1177/2056305120948250.

Ward-Griffin, Catherine, and Patricia McKeever. 2000. "Relationships between Nurses and Family Caregivers: Partners in Care?" *Advances in Nursing Science* 22 (3): 89–103. https://doi.org/10.1097/00012272-200003000-00008.

Weizenbaum, Joseph. 1976. *Computer Power and Human Reason: From Judgment to Calculation.* San Francisco: W. H. Freeman & Co.

Part 3

POSTHUMAN FICTIONS, FUTURES, AND BODIES

Chapter 10

Im/Material Bodies

Queering Embodiment through Performance Art and Technology

Joep Bouma

At the intersection of gender, new media technologies, and art, this chapter situates the conceptual parallels that transgender and queer studies have with posthumanism,[1] specifically as they impact processes of embodiment. Studying the consequences and possibilities of such fields as robotics, biotechnological enhancement, and artificial intelligence for intersectional issues of social justice, the unique position that transgender subjects hold within these dynamics requires scholarly attention. As technology becomes enmeshed into culture through processes of digitization and automation, new questions around transgender justice arise, ranging from pragmatic to conceptual matters and from collective needs to individual experiences. Departing from a notion of technology as potentiality for transgender subjects, artistic practices, particularly, constitute a valuable field of inquiry inside this larger endeavor. Within the arts, the creative, aesthetic, and communicative qualities of new technologies have been explored by queer and trans artists. Through their works, they demonstrate the possibility of new modes of gender embodiment through intimate engagement with the digital.

Two central strands of discourse synthesized here relate primarily to theorizations of embodiment, on the one hand, as a reconfiguration of gender and its dependency on bodies, and on the other, as it transforms under increasingly pervasive technologies in our everyday lives. Technologies, importantly, are not considered as a neutral given or as inherently progressive, but as O'Riordan states, "as objects and discourses, material and symbolic, imaginary and actual," thus allowing forms of intervention (2017, 3). The creative uses of digital technologies by trans and queer artists, then, contribute to the advancement of these objects and discourses and should be explored and situated. Combined with decades of academic efforts, these developments culminate into a queer posthumanism in which the relationship between trans people and technology is continuously materializing its potentiality.

This is where I propose including another intersection: that between gender theory's conceptualizations of performativity and performance art. A strong tradition exists within feminist performance art, and the medium of bodies on stage has known many great binary trans artists[2] as well as queer artists with other gender identities. Over time, as new technologies developed and became available, several such artists have incorporated and advanced the potentiality of technological innovations into their artistic explorations and statements. Whether objectives of their practices have been political, conceptual, or personal in nature, we can speak of an oeuvre along these thematic and methodological lines.

This chapter looks to contemporary artists with distinct outlooks on both gender and technology, to assess the directions these works point toward. Taking neither gender nor technology as a given, what possibilities do these artists manifest? The core goal of this chapter is to recognize and situate the academic value of trans and queer artists' practices. It does so by analyzing the following works: the album *OIL OF EVERY PEARL'S UN-INSIDES*, by SOPHIE (2018); the instrument *Music Dildo*, created and used in performances by Tami T (2016); the performance pieces *The New Body,* by Astrit Ismaili (2018), and *There Are Certain Facts That Cannot Be Disputed*, by Juliana Huxtable (2015); and the *My Sex* video work, by Pastelae (Jonsson 2018). Their different understandings of gender embodiment, as presented in their works, highlight process and inquiry, as opposed to fact and conclusion. These works' collective proximity to popular culture is valuable, as it demonstrates their particular engagement with the hegemonic and makes their innovation accessible outside of the confines of academia and white cube art industries.

Building on important works in queer studies, performance and theater studies, new media philosophy, and feminist posthumanism, what follows is an outline of the aforementioned discursive developments. Identifying theoretical contributions on the queering of embodiment and the mediation and performance of bodies, I develop a framework of discursivity that examines the role of imaginaries, practices of becoming, and technological mediation within these practices. An analysis of the works then shows how an expansion into artistic practices constitutes an advancement of queer posthumanism that is necessary for its progress and its relevance to trans and queer subjects.

QUEERING BODIES

"Explor[ing] the encounter between flesh and discourse" (MacCormack 2009, 112), various works of queer theory have greatly contributed to our understanding of gender embodiment. Building on Judith Butler's *Bodies That Matter*, in which she stresses that "sex" is "not a simple fact or static condition of a body" (1993, 1), Gayle Salamon's *Assuming A Body: Transgender Rhetorics of Materiality* (2010) incorporates psychoanalysis, phenomenology, and queer theory to further formulate the processes of gender embodiment. In particular, she skillfully clarifies aspects that tend to be misinterpreted in exchanges between essentialism and constructionism: a contentious dualism seems to persist between the realms of the "Real" and the "Imaginary," in which the latter is quickly misread as "synonymous with imagination, hallucination, delusion, or conjecture and [as] fundamentally opposed to materiality and reality (which are functionally synonymous)" (35). Instead, the book argues that, while it is not disputed that "the body is a material structure" (88), bodily coherence, or the anatomical "truth" of our identities, only exists "at the level of the imaginary, which both governs the production of that anatomy and makes it available to the psyche" (35). Fantasy, here, is not an inconsequential activity of the subject but rather "something that enables the subject" (35). Acknowledging the central function of the Imaginary realm allocates an additional dimension of significance to the works created by trans and queer artists, as they forefront it and make our engagement with it explicit and inevitable.

Working with the Imaginary, the relevance of the image—of the visual mediation of bodies—is evident. Schilder's work on body image assumes its definition to be "both representative and constitutive of the self" (quoted in Salamon 2010, 26) and to be flexible as well as social. This understanding of the visual is radically different than the essentialist conception of a visual exterior covering or hiding the Real that lies beneath a surface of performativity. Rather, Bates (2015) formulates bodies as emergent through relations between parts, powers, properties, capabilities, liabilities, and dispositions (140), as a result of which "humans both have and are a body" (138). Two elements to be taken away here are the fragmented and unfinished nature of embodiment and the understanding of the body as a verb rather than a noun. An emancipatory quality of the queer theory notion of moving from queer*ness* into queer*ing* is reflected here and informs the selected artworks.

Specifying the elements comprising this process, Schilder (1950) points toward pain, itching and erogenic zones, but also actions of others toward our bodies as well as our own touch. Additionally, Salamon (2010) identifies image, posture, and touch as enabling the *corps propre*, referencing

Merleau-Ponty. Butler (1993), finally, posits language as key in this process, "for language both is and refers to that which is material, and what is material never fully escapes from the process by which it is signified" (68). Indeed, "What the 'real' body tells us—or, rather, what it silently displays, without benefit of language—is nothing" (Salamon 2010, 88). This leaves us with a regard for visual, textual, and tactile media as being inseparably intertwined with our own bodies. As such, "[c]ommunications technologies and biotechnologies are the crucial tools *recrafting* our bodies," Haraway argues (1991, 302, emphasis added). What it means for a posthuman subjectivity when these media are digitized is addressed below.

Having located several facets of embodiment processes, a degree of agency must be recognized, calling for a "narcissistic investment in one's own flesh" (Salamon 2010, 41). Rather than passively undergoing the consequences of the language, images, and physical sensations one is confronted with, we are in fact dealing with a process of creation, of poetry, of potentiality. Gender is conceived of as not one out of two structures imposed upon us, but, as Feinberg notes, "the poetry each of us makes out of the language we are taught" (quoted in Salamon 2010, 73). Accordingly, Bates positions our body as a potentiality that is not determined but rather elaborated by culture (2015, 141).

The relationality of embodiment is stressed by Salamon, as bodies "function as a mediating entity between self and world" (2010, 31). She investigates relationality through a nuanced re-emphasis on the "sexual schema," asking: "What might it mean to suggest that the body itself comes to be though desire?" (46). MacCormack (2009) further queers relationality by introducing seduction as a technique of becoming, one that I will incorporate into the analysis of the selected artworks:

> By exploring the pure possibility of bodies, always in excess of what non-perverts are allowed to do, we can be seduced by the risks and delights of not knowing what will happen [. . .]. Becoming shares with seduction the entering into an alliance with an entity by which we are seduced because it is other, because it is strange and not us. Seduction is not a desire to know or assimilate the other, it wants the other to change us and us the change the quality of the other to create a unique hybrid beyond any sexual narrative. Seduction is a sexual technique of queer becomings. (115)

Formulating these processes as such does not automatically alleviate the discomfort and dysphoria of embodied becoming. Incompleteness and fragmentation need not result in a perpetually unfulfilling hopelessness, but even if they do, they can constitute a viable and constructive modus operandi. Speaking, moving, touching, and visualizing through desire and creativity, then, it is the processes of assuming trans and queer bodies that the artists are engaging in and offering insight into.

MEDIATING BODIES

A parallel to bodies as "incoherent conglomeration[s]" of "insignificant elements" (Silverman 1996, 22)—or *bits*—presents itself when expanded into the digital: "Microbes and bits are *both* media of existence" (Peters 2015, 52, emphasis added). An important conceptual lineage of posthumanism and cyberfeminism, pioneered by Donna Haraway and Katherine Hayles, is of great importance here. A variety of theoretical approaches has been offered, including the notions that the ongoing "disembodiment of information" has already constructed us into cyborgs (Hayles 1999); that such machine–organism hybrids are both a social reality and a productive fiction (Haraway 1991, 291); and that, given the incompleteness of bodies, they are co-produced through "interface with the computers, forming part of an information machine in which the body's limbs and organs will become interchangeable parts with the computer and with the technologization of production" (Grosz 1992, 252).

Braidotti (2013) identifies the intimate connection to trans embodiment, observing transness as "a dominant posthuman *topos*" as machines are "neutralized as figures of mixity, hybridity and interconnectiveness" (97). Halberstam (2005) goes as far as to equate expanded technotopian and transgender embodiment entirely, referencing the inclusion of new organs and the expansion of sensoria (101).[3] The pitfall of technomania, however, would be an assumption of posthuman subjects as beyond racialized or sexual difference (Braidotti 2013, 97). Starkly moving away from the notion of disembodiment, Hansen develops a phenomenology in which bodies under media convergence actually increase in centrality, that is, as framers of information (2004).

While posthumanism's conceptual heritage is as valuable as it is fascinating, Rowan (2015) is not alone in the conviction that "these conversations too often revolve around the metaphysically inflated phantom of Technology as such rather than engaging the specific ways in which particular technologies are put to use for certain ends within distinct social assemblages" (8). Turning to a more pragmatic subjectivity, Lanier (2010) suggests looking at media designs specifically, and in what directions they stimulate the identified potentiality to develop, while also emphasizing our agency within this development (5).

Whereas the technological possibilities for trans people in the medical field are largely in the hands of medical experts, one could argue that artistic practices allow for more autonomy and access in recrafting our queer bodies with(in) the digital. Looking at various *techniques* of employing posthumanism allows us to further understand and expand their potentialities. More than a concept, less than a substitute, queer digital being can use the posthuman as a navigational tool (Braidotti 2017, 40). Using readily available technological products that in themselves do not consider the specificities of trans and queer lives, artists can queer commercial, public spaces. Additionally, building technological bodily expansions for a performance, even if they are not wearable in daily life, further enables the "creation of new bodies in an aesthetic realm" (Halberstam 2005, 103). The incorporation of radically de- and reconstructed bodies into our digital mediascapes, offering new possibilities that cover but are not limited to gender, constitutes a valuable expansion of our accessible imaginaries.

This further points to a dimension of shared or communal becomings, in which compassionate interdependence, as Braidotti phrases it, is fully acknowledged. She culminates a contemporary posthuman ethics that sacrifices "the fantasy of unity, totality and one-ness, but also the master narratives of primordial loss, incommensurable lack and irreparable separation" (2013, 100). As has been established, relationality alongside or through desire was found equally inherent to processes of bodies and gender. This parallel aids the localization of the specific arenas in which queer posthumanism can come to fruition. Embracing the premise of *not-One*, Braidotti argues, "anchors the subject in an ethical bond to alterity, to the multiple and external others that are constitutive of that entity which, out of laziness and habit, we call the 'self'" (2013, 100). Investing in an inclusive and constructive relationality, this bond to alterity is strengthened through artistic practice.

PERFORMING BODIES

Recognizing the act of creation that is required in queer posthuman becoming, the artist's work is vital. Formulating "pure dislocations of identities" through perversion (Braidotti 2013, 99) is in itself an artistic process, and these constructions are inescapably implicated in the advancement of queer futures. Putting bodies center stage in performance, specifically, allows the artist's body to investigate the in/abilities of embodied subjectivity, "render[ing] palpable possibilities for unanticipated signification" (Reinelt 2002, 213). It is this ability to experience that art as a method specifically enables, as opposed to what theory alone provides.

Forefronting bodily interaction with technology in performance signifies a move away from using bodies as theatrical tools to reveal inner processes, from the false distinction between inner truths and deceptive surfaces. Actively intervening with bodies in performance, rather, assumes bodies as their own theaters of change: "the flesh serves as the threshold upon which subjectivity is produced through its interaction with technology" (Case 2002, 43). A lineage of feminist performance art teaches us about what Schneider (1997) dubs the "explicit body," in which women use their bodies "beside themselves, [. . .] as a means of making explicit the historical staging of that body" (176).

Whereas a previous generation of trans artists, continuing this tradition, often focused on their physical transition, however cathartic, more recent work by trans artists explores other social, emotional, and physical complexities (Heinlein 2016). Explicit bodies, when trans people are concerned, quickly risk rearticulating "a fixation with 'the' surgery" (Salamon 2010, 113) and reinstating those narratives of lack, loss, and separation that Braidotti (2013) pleas to break. Bodies in art, constructively, have seen a transformation parallel to the expansion of posthumanist discourse, for example, through what Jones (1998) calls dispersed subjects: "younger artists tend to explore the body/self as technologized, specifically *un*natural, and fundamentally unfixable in identity or subjective/objective meaning in the world" (199). As a result, art has become a way to expand and materialize the realms of possibility in queering posthuman embodiment.

Further developing such terms—dislocated, unnatural, unfixable—into a framework, Muñoz (1999) offers disidentification as a specific mode of performance, built by queer artists of color. It makes positionality toward dominant ideology explicit, opting neither for assimilation nor for strict opposition (11), but transforming it from within through reconstructing the encoded meanings in its cultural objects (31). This is where the tangible immediacy with pop culture in the selected artworks becomes significant: "disidentification is a step further than cracking open the code of the majority; it proceeds to use this code as raw material for representing a disempowered politics or positionality that has been rendered unthinkable by the dominant culture" (31). Common to such practices are the sensations of uneasiness and of desire, often appearing hand-in-hand, as they are portrayed by the performer and/or experienced by their audiences. The uneasiness of illegibility, paired with the desire for the alleviation of essentialism's categorical restraint, draws us into the productive realm of potentiality. Operating through MacCormack's (2009) technique of seduction, then, these artists mobilize the textual, visual, and tactile media that we are entwined with to create anew. I will now elaborate on the contributions to this framework of mediated queer embodiment that the selected artworks synthesize.

SOPHIE

No stranger to manifesting pleasure, electronic musician SOPHIE's work perhaps provides the most literal exemplification of the developments outlined above, in that it mobilizes language, images, and sounds directly connected to cyborg imaginaries, gender performativity, and popular culture. Her 2018 debut album *OIL OF EVERY PEARL'S UN-INSIDES* and its accompanying visuals give a face—albeit everchanging—to her disorientating music. Her first full on-camera appearance in *It's Okay To Cry* (SOPHIE and Harwood 2017) uses dramatic cheekbone prosthetics, while *Faceshopping* (SOPHIE and Chan 2018; https://www.youtube.com/watch?v=es9-P1SOeHU) shows an animated rendering that annexes her face, twisting and splitting it into endless iterations. As the most animate bodily feature, the face has a special place in performance, "because of its essential futurity," defying static objecthood, visualizing our ability of transcendence (Sartre quoted in Erickson 1990, 236), which is echoed in *Faceshopping*'s lyrics.

In a flash between these facial shots, the common sight of an "I'm not a robot" Captcha appears on the screen. Through assemblage, language, and imagery, SOPHIE synthesizes the real, claims artificiality and virtuality as legitimate tools of embodiment in a digitally mediated world, and allows her sonic and visual creation to exist perhaps as constituents of her body. Grounded in the flesh, malleable in appearance, evading signification, yet always real, this face can be seen as "*both* separate from *and* relative to herself in her general struggle with historical legacies of disembodiment: paradoxical and impossible as she is, being, herself, the previously unimaginable" (Schneider 1997, 184; italics in original). The disembodied voice, further liberated from restraints and gendered demarcation through means of digital vocal modulation (Blanchard 2018), adds to the dispersion of the subject in her work, as does the backlighting in her live shows that presents the artist as a barely identifiable contour.

Demanding reconsideration of dualisms as real and fake, human and machine, organic and synthetic, SOPHIE states "What *is* real? Being trans," and, importantly, her transness goes beyond gender (Juzwiak 2018; italics in original). Playing with discursivity in dis/embodiment, and the potentiality of it, she conjoins both in the song title "Immaterial" (2018). Indeed, SOPHIE's aesthetic builds heavily on the lasting appeal of queer posthumanism, expanding it into contemporary popular technoculture and granting it palpability. Through bringing her audience as close to this state as she can; through accrediting the technologies around us with the ability to bring us there as long as we perceive it as indeed Real; through forwarding the trans "spirit of self-engineering by using sound, [image, and language] as artistic flesh for surgical operation" (Blanchard 2018). "The only way I can put it," she proclaims herself, "is having fun in your body" (Ravens 2018).

JULIANA HUXTABLE

Creating through language as well as through self-portraiture and performance, Huxtable's work deals with her relationship to her body as a prerequisite to create, as well as a source of relational tension. Working to undo the skewed institutional narratives her body is put into, as well as negotiating the dynamics of exchange that come with performing for an audience, a rather different approach to the engagement with posthumanism is chosen, through a framing focusing on information and access (Chiaverina 2015). References to common online experiences, current and of recent history, acknowledge the existential aspects of quotidian technologies as our lives become entwined with them. Proceeding to use this framework to address the precarity of her social reality, she renders her body as information, relationality as exchange of information, and identity trouble and historical marginalization as issues of access and data visualization.

Actively engaging with early utopian visions of the Internet age in the poetic performance *There Are Certain Facts That Cannot Be Disputed* (http s://www.youtube.com/watch?v=2W2hpfKrtu4), Huxtable exposes the ways in which human error and distorted relations of power have expanded into the digital. As Losh and Wernimont state in *Bodies of Information*, a problem in shared digital knowledge arises "if fundamental infrastructures of information are designed solely to sort data into binary or mutually exclusive categories" (2018, xviii). Citing *Encyclopedia Africana*, an ambitious but discontinued endeavor of Afrocentric history-making through commons-based peer production that perished in competition with Wikipedia, Huxtable voices frustration with the Internet's "transition to something like an oligarchy" (Burns 2015). This frustration is echoed as she places a repeated but unrequited plea to an ex-lover Geo to give her her texts back. Emphasizing the "material, situated, contingent, tacit, embodied, affective, labor-intensive, and political characteristics of digital archives [. . .] rather than friction-free visions of pure Cartesian 'virtual reality' or 'cyberspace'" (Losh and Wernimont 2018, xi), Huxtable describes how a history of her body is being assembled by a handful of presumably white men through static artifacts that lie locked away. The visual aspect of embodiment is referenced by detailing the politics of data visualization and its direct impact on bodies, including her own.

The relationality that has been offered as inherent to assuming a body is thus problematized by investigating the historical skewedness of those relations. In a powerful gesture, the performance realizes canonical history as a produced fantasy, subsequently granting legitimacy to the creation of alternative fantasies. This possibility is addressed through visual mediation in the final sequence, in which the sense of freedom and playfulness that Internet visual culture enables is explored: "It's shallow on the one hand but it's more powerful because it's more accessible, it's more fun, but it also from the jump acknowledges history as something that should be played with, and it's always a question of fantasy," she explains (quoted in Chiaverina 2015). Creating new relations between bodies, cultural texts, and historical archives, Huxtable mobilizes media art's aesthetic of assemblage to offer a highly personal work of disidentification and digital embodiment.

TAMI T

Highlighting the significance of touch as a mediator of our bodies, and its potential as a site of intervention, musician Tami T has created and programmed a tactile digital instrument (2019b). Referred to as a musical strap-on dildo (https://www.instagram.com/p/BvrnhcGhB2b/), she attaches it to her crotch with rope and produces synthesized sounds by stroking and hitting it with a drumstick. Writing on the morphological imaginary, Butler wrote: "to speak of the lesbian phallus as a possible site of desire is not to refer to an *imaginary* identification and/or desire that can be measured against a *real* one; on the contrary, it is simply to promote an alternative *imaginary* to a hegemonic imaginary and to show, through that assertion, the ways in which the hegemonic imaginary constitutes itself through the naturalization of an exclusionary heterosexual morphology" (1993, 92; italics in original). Tami T's trans femme perspective presented with the instrument further disrupts any predominant narrative of transfemininity and obstructs the confinement of her morphology to dominant forms. Reflected in the song "It's Not Your Right To Know" (2019a),[4] Tami T premeditates the anatomical fixity imposed on trans people and replaces any invasive and misguided inquiries into "what really lies beneath" by an encounter with an explicit body of her own making.

Witnessing her perform with the instrument, consequently, evokes a sense of an intimacy that is not voyeuristic but rather proposes a safe, shared exploration of the vulnerabilities of our bodies aided by technology. The themes of care and community recur in Tami T's lyrics, too (2019a). Acknowledging the precariousness of her community's lives, these lyrics cover queer sexual pleasures ("Face Riding") as well as perils of harassment and violence ("So Afraid"). This enhances Tami T's work in also accounting for the lived experiences that exist in relationality, outside of what is created and presented on stage. Disidentification, indeed, "is not to pick and choose what one takes out of an identification [. . .]. Rather, it is the reworking of those energies that do not elide the 'harmful' or contradictory components of an identity. It is an acceptance of the necessary interjection that has occurred in such situations" (Muñoz 1999, 12).

The overt assemblage of her body provokes a *binary terror*, suggesting "a collapse of the space between phantomic appearance and literal reality, interrogating the habitual ease of our cultural distinctions" (Schneider 1997, 99). The compelling playfulness and simplicity of Tami T's aesthetic choices in this provocation and the movement away from the purely "virtual" make her work all the more powerful, synthesizing a reality that is welcoming trans and queer audiences to join and *become* through.

BROOKE CANDY AND PASTELAE

3D illustrator Pastelae created a video work for musician Brooke Candy's collaboration with queer rapper Mykki Blanco, art and activist collective Pussy Riot, and singer-songwriter MNDR, called *My Sex* (Jonsson 2018; https://www.youtube.com/watch?v=hxmvI3ECcXQ). Within a contemporary aesthetic of pastel-colored, glossy, hybrid figures that constitute a cyborg imaginary for the Instagram era, the result proposes a morphology of fluidity. The video starts with an introduction to the participating musicians as floating animated characters. Similar to SOPHIE's *Faceshopping* video, the figures show direct linkages to Candy's and Blanco's bodies: they are stamped with reproductions of their respective tattooed skins. Rather than portraying exact replicas, however, the bodies in the video start multiplying, deflating, and eventually morphing. Candy's skin is applied to Blanco's posture, and vice versa. The boundaries between the artists' bodies are negated through seduction, resulting in a new representation of a non-binary embodiment. This representation continues a framing of bodies as canvases, emphasizing the ways in which they can be constructed and altered, not only through tattoos but also through the animated application of another skin. Augmented by the absence of legible facial expressions, these bodies are no longer surfaces that cover interior subjectivities but rather surfaces on which subjectivities are projected (Salamon 2010, 22).

Staging this act of projection, an interesting outlook on autonomy is emphasized. On the one hand, the work demonstrates a clear link to Braidotti's (2013) non-unitary subject, visualizing a literal *interconnectiveness* of bodies. On the other hand, sexual empowerment is proclaimed through repeatedly emphasizing individual ownership over one's body, outlining what "my own sex" is and is not. It can be argued that what the video brings to life is Braidotti's (2013) experiment with the posthuman, as well as Salamon's

(2010) sexual schema. "We need to experiment with resistance and intensity in order to find out what posthuman bodies can do," the former proposes, in order to "rediscover the notion of the sexual complexity that marks sexuality in its human and posthuman forms" (Braidotti 2013, 99). Sharing bodies without losing ownership, through queer practices and visualization of sexuality, contributes to reestablishing these complexities in the accessible realm of contemporary online visual culture. Investigating the potential of sexuality in understanding our relational subjectivity, Salamon finds that it posits a "means by which a transformation from ideality to particularity becomes possible" (2010, 57). The unsettling, interconnected autonomy presented in the video is expressed, in that "my embodied existence [is] held in this inescapable and tensile paradox: I am for me and I am for the other, and each of these modes of existence realizes itself in my body" (2010, 57).

ASTRIT ISMAILI

Further broadening the scope of bodily investigation beyond gender, Ismaili's *The New Body* (2018) invents new mediations and significations of their bodily histories. Shifting our attention to what it might mean to *sonify* bodies, the performance entails the artist standing on a rotating platform in a dark or dimly lit room, wearing a type of technological armor that encompasses their hips, head, hands, and feet. The device, similar to Tami T's strap-on, can be seen as an instrument but rather than being operated by touch, it responds to posture and movement, tightening its relation to the artist's body. The result is productive; by abstracting the body into sound the artist transcends its conventional readings into new perspectives on the human body and mind. The scale of the instrument makes it almost into an entire new bodily shell, obscuring the skin as the external delimitator of the wearer's body. By responding to every gesture, the instrument first seems to speak on behalf of the subject, but then their "own" voice joins and sings with the voices coming from the instrument. The performance presents an intimate bond that is concerned with new becomings more so than with making their body legible to an audience. Defending a right to remain undefined, Ismaili upsets repressive social classification.

Rather than escaping their flesh, however, Ismaili honors the personal and shared histories that it contains: "it's about wanting to change without losing all the knowledge captured from living" (Ismaili 2019, personal communication, October 16). Echoing the central theme of histories in Huxtable's work, *The New Body* seeks modes of queer posthuman subjectivity that do not elide the physical or the historical but rather work to establish instrumentality in how they get to determine our bodies. "I want to start over, not from zero, but from now," Ismaili proclaims at the end of their performance (2018).

The introduction of choral traditions of multivocality into this new body restates the importance of queer community, solidarity, and autonomy (Ismaili 2019). With the audible presence of *others* in every move Ismaili makes, their body becomes an expression of not-One. The fashioning of a new body hence does not constitute an "entirely voluntaristic project, somehow freely chosen by the subject" (Salamon 2010, 31), but presents an embodiment that forefronts Braidotti's compassionate interdependence (2013, 100). A perspective on technological possibilities for queer subjects is offered that acknowledges the centrality of bodies and proposes a valuable outlook on autonomy and relationality.

QUEERING POSTHUMAN
EMBODIMENT THROUGH ART

Without compromising the corporeality of gender and sexuality, the works analyzed present a technological incorporation that does not dwell in futile questions of real or unreal. Instead, an alternative imaginary is asserted, one that defies conclusions regarding mutual relations between bodies, genders, and technologies. In their respective efforts, they work through different degrees of digitization, using different media to investigate different techniques of queering embodiment. Identifying commonalities, however, we can first observe a variety of dispersed bodies that are newly created across aesthetic realms, through image, language, sound, and touch. Popular culture, here, functions as a site in which new imaginaries that enable subjects are constructed through disidentification. Rather than denying *materiality* to their bodies, the works make it malleable through the labor of disidentification and technocultural innovation. In doing so, they offer different levels of fluidity in their conceptions of boundaries between the digital and physical, the singular and the multiple. Incorporating the conceptual lineage of queer posthumanism, the works are found to present embodiment as constructively instable and relational.

The hybridity foregrounded carries a specific power that should be addressed, for the term in itself could cover assimilation, mimicry, as well as "creative transcendence" (Shohat quoted in Muñoz 1999, 78). Belonging to the latter, the hybridity in the artworks' bodies establishes a constructive instability. What this instability points out is the processual nature of embodiment. The *becoming* that is presented is not one of *turning into* (MacCormack 2009, 121), as the artists provide no finite alternative in replacement of existing stagnant categories. What they convey, as practices of queering posthuman embodiment, is an open-ended, intentional alliance with technology and with multiple and external others. The propositions found in literature on forging this alliance through principles of seduction and care can be traced in the prevalent themes of sexuality and multiplicity. The notion of pleasure through the expansion of our im/material bodies and the blurring of its limits allows us to be seduced into a becoming that is shared: not as a romantic abstraction of intimacy through sacrifice, nor as an *ersatz* substitute for flesh but as a reconsideration of a Real that is not-One. A key contribution lies in taking the trans activist objective of "redefining realness" (Mock 2014) and expanding it into updated iterations of posthuman imaginaries. The instability of queer posthuman subjectivity is convincingly constructive, because its *realness* is explicitly stressed even as it is being redefined.

What we can lastly discern is a gesture toward specifying the ways in which technology can be mobilized. These artists make the potentialities of queer, mediated bodies available through audiovisual, tactile, and textual means. In this sense, the selected artworks are part of a praxis that gives substance to a "materialist politics of posthuman differences [which] works by potential becomings that call for actualization" (Braidotti 2013, 100). Aware that technological "progress" is in fact not inherently progressive, as becomes painfully clear in Huxtable's work, the works fulfill an urgent need for queer technological innovation. For queer subjects, having access to the bodily realities crafted here can be of great significance within the everyday existential navigation of assuming a body that is already increasingly digital. The technologies that have informed and have been formed by these artists, then, demand consideration by art critics and new media scholars alike, as they mobilize and expand the lineage of queer cyborgs into new realities. Finally, it should be stressed that these endeavors, both artistic and academic, should keep foregrounding the aim of freeing queer subjects in their practices of embodiment. Ultimately; "It's fascinating to talk about this stuff, but it's more fun to dance" (SOPHIE quoted in Juzwiak 2018).

POSTSCRIPT

This chapter is dedicated to SOPHIE, who passed away after a tragic accident on January 30, 2021. The access to playful and empowered futures that her work has granted us, however, remains tangible. As we take it upon us to honor and expand on her contributions, SOPHIE enters a new realm of im/materiality.

NOTES

1. For further discussion of posthumanism, see Lakshmanan, chapter 6, this book. — Ed.
2. By "binary trans artists" I refer to trans people who *do* identify within the gender binary (i.e., trans men or trans women). I make this distinction to indicate

that the medium has been explored by a wide variety of trans artists, both binary and non-binary/other.

3. An interesting corollary to this view is finding techno-utopian frames even within techno-dystopian creative works. For example, "queer erotechnics" (see van Veen, chapter 12, this book) may be found in the otherwise-panned MMORPG *Cyberpunk 2077*. As game journalist Riley MacLeod (2020) notes: "The character creator gave me something no other creator has: instead of locking me into 'male' or 'female,' it let me choose a traditionally masculine body type and voice, then give my character a vagina, making my character undeniably a trans man, like me." — Ed.

4. This and the subsequent referenced songs are all from her album *High Pitched and Moist* (2019b).

REFERENCES

Bates, Stephen R. 2015. "The Emergent Body: Marxism, Critical Realism and the Corporeal in Contemporary Capitalist Society." *Global Society* 29 (1): 128–47. doi :10.1080/13600826.2014.974514.

Blanchard, Sessi Kuwabara. 2018. "How SOPHIE and Other Trans Musicians Are Using Vocal Modulation to Explore Gender." *Pitchfork* (blog). June 28, 2018. https://pitchfork.com/thepitch/how-sophie-and-other-trans-musicians-are-using -vocal-modulation-to-explore-gender.

Braidotti, Rosi. 2013. *The Posthuman.* Cambridge: Polity Press.

Braidotti, Rosi. 2017. "Four Theses on Posthuman Feminism." In *Anthropocene Feminism*, 21–48. 21st Century Studies. Minneapolis: University of Minnesota Press.

Burns, Charlotte. 2015. "Juliana Huxtable Interrogates 'Older, Whiter Versions' of History at MoMA." *The Guardian*, November 6, 2015, sec. Art and design. https://www.theguardian.com/artanddesign/2015/nov/06/juliana-huxtable-moma -performa-15-new-york.

Butler, Judith. 1993. *Bodies That Matter: On the Discursive Limits of "Sex."* New York: Routledge.

Case, Sue-Ellen. 2002. "Performing the Cyberbody on the Transnational Stage." *Gramma: Journal of Theory and Criticism* 10: 41–57. doi:10.26262/gramma. v10i0.7254.

Chiaverina, John. 2015. "'It's Always a Question of Fantasy': Juliana Huxtable On Her Indisputably Brilliant Performa Commission." *ARTnews* (blog). November 20, 2015. http://www.artnews.com/2015/11/20/its-always-a-question-of-fantasy -juliana-huxtable-on-her-indisputably-brilliant-performa-commission.

Erickson, Jon. 1990. "The Body as the Object of Modern Performance." *Journal of Dramatic Theory and Criticism* 5 (1): 231–46.

Grosz, Elizabeth. 1992. "Bodies-Cities." In *Sexuality & Space*, edited by Beatriz Colomina, 241–53. Princeton Papers on Architecture. New York: Princeton Architectural Press.

Halberstam, Judith. 2005. *In a Queer Time and Place: Transgender Bodies, Subcultural Lives.* New York: New York University Press.

Hansen, Mark B. N. 2004. *New Philosophy for New Media*. Cambridge: MIT Press.

Haraway, Donna. 1991. "A Cyborg Manifesto: Science, Technology and Socialist-Feminism in the Late Twentieth Century." In *The Cybercultures Reader*, edited by David Bell and Barbara M. Kennedy (2001), 291–324. New York: Routledge.

Hayles, N. Katherine. 1999. *How We Became Posthuman: Virtual Bodies in Cybernetics, Literature, and Informatics*. Chicago: University of Chicago Press.

Heinlein, Sabine. 2016. "The Transgender Body in Art: Finding Visibility 'in Difficult Times like These.'" *The Guardian*, November 18, 2016, sec. Art and design. https://www.theguardian.com/artanddesign/2016/nov/18/transgender-art -trans-hirstory-in-99-objects.

Huxtable, Juliana. 2016. *There Are Certain Facts That Cannot Be Disputed*. Performance. New York, Museum of Modern Art. https://www.youtube.com /watch?v=2W2hpfKrtu4.

Ismaili, Astrit. 2018. *The New Body*. Performance. Prishtina, LambdaLambdaLambda.

Ismaili, Astrit (@astritismaili). 2019. "Tonight at @korzotheater . . . ." *Instagram*, September 14, 2019. https://www.instagram.com/p/B2ZTtxho_WD.

Jones, Amelia. 1998. *Body Art/Performing the Subject*. Minneapolis: University of Minnesota Press.

Jonsson, Josefin. 2018. *My Sex*. ManyVids & Freaky Princess Inc. https://www.you tube.com/watch?v=hxmvI3ECcXQ

Juzwiak, Rich. 2018. "SOPHIE on Her New Album, Old Disco, and Expressing Trans Identity in Music." *Jezebel* (blog). June 15, 2018. https://themuse.jezebel.com /sophie-on-her-new-album-old-disco-and-expressing-tran-1826863700.

Lanier, Jaron. 2010. *You Are Not A Gadget: A Manifesto*. London: Allen Lane.

Losh, Elizabeth, and Jacqueline Wernimont. 2018. *Bodies of Information: Intersectional Feminism and the Digital Humanities*. Debates in the Digital Humanities. Minneapolis: University of Minnesota Press. doi:10.5749 /j.ctv9hj9r9.

MacLeod, Riley. 2020. "30 Hours with *Cyberpunk 2077* Brings Mixed Feelings." *Kotaku*, December 7, 2020. https://kotaku.com/30-hours-with-cyberpunk-2077 -brings-mixed-feelings-1845824151?utm_campaign=Gizmodo&utm_content=1 607378027&utm_medium=SocialMarketing&utm_source=facebook&fbclid=I wAR1n_nVTf2KFztwPyh8valRRuucY-VUwXHAbbttCY0i1yZixcdyKuAYJRrw.

MacCormack, Patricia. 2009. "Queer Posthumanism: Cyborgs, Animals, Monsters, Perverts." In *The Ashgate Research Companion to Queer Theory*, edited by Noreen Giffney and Michael O'Rourke, 111–26. Queer Interventions. Farnham: Ashgate Publishing Limited.

Mock, Janet. 2014. *Redefining Realness: My Path to Womanhood, Identity, Love & So Much More*. New York: Simon & Schuster, Inc.

Muñoz, José Esteban. 1999. *Disidentifications: Queers of Color and the Performance of Politics*. Vol. 2. 3 vols. Cultural Studies of the Americas. Minneapolis: University of Minnesota Press.

O'Riordan, Kate. 2017. *Unreal Objects: Digital Materialities, Technoscientific Projects and Political Realities*. London: Pluto Press.

Peters, John Durham. 2015. *The Marvelous Clouds: Toward a Philosophy of Elemental Media*. Chicago: University of Chicago Press.

Ravens, Chal. 2018. "SOPHIE: Earthly Pleasures." *Crack Magazine*, May 1, 2018. https://crackmagazine.net/article/long-reads/sophie-earthly-pleasures.

Reinelt, Janelle G. 2002. "The Politics of Discourse: Performativity Meets Theatricality." *SubStance* 31 (2): 201–15. doi:10.1353/sub.2002.0037.

Rowan, Rory. 2015. "Extinction as Usual?: Geo-Social Futures and Left Optimism." *E-Flux Journal*, no. 65 (May). http://supercommunity.e-flux.com/texts/extinction-as-usual-geo-social-futures-and-left-optimism/.

Salamon, Gayle. 2010. *Assuming A Body: Transgender and Rhetorics of Materiality*. New York: Columbia University Press.

Schilder, Paul. 1950. *The Image and Appearance of the Human Body: Studies in the Constructive Energies of the Psyche*. Vol. 10. Physiological Psychology. Abingdon: Routledge.

Schneider, Rebecca. 1997. *The Explicit Body in Performance*. New York: Routledge.

Silverman, Kaja. 1996. *The Threshold of the Visible World*. New York: Routledge.

SOPHIE. 2018. *OIL OF EVERY PEARL'S UN-INSIDES*. London: Transgressive Records.

SOPHIE, and Aaron Chan. 2018. *Faceshopping*. Los Angeles: Industry Plant. https://www.youtube.com/watch?v=es9-P1SOeHU.

SOPHIE, and Nicholas Harwood. 2017. *It's Okay To Cry*. Los Angeles: Ways & Means. https://www.youtube.com/watch?v=m_S0qCeA-pc.

Tami T. 2019a. *High Pitched and Moist*. Berlin: Trannytone Records.

Tami T (@_tami_t_). 2019b. "'My Favorite Instrument." *Instagram*, March 31, 2019. https://www.instagram.com/p/BvrnhcGhB2b/.

Chapter 11

Estranged World

Tenets of Xenofeminism and Tropes of Automated Alienation in Contemporary Alien Films

Christopher M. Cox

This chapter undertakes a textual analysis of recent extraterrestrial alien films—*Arrival* (2016), *Alien: Covenant* (2017), and *Annihilation* (2018)—to interrogate the complications of alienation as a central facet of xenofeminist techno-politics and the risks automated technologies pose to aims of xenofeminist alienation.

Xenofeminism is a recent strand of techno-feminism predicated on the pursuit of gender abolition, antiessentialism, and techno-materialism (Laboria Cuboniks 2015; Hester 2018). In the case of the latter, xenofeminism's techno-materialism situates technology as a significant "sphere of activist intervention [. . .] capable of acting as vectors for new utopias" in which utopian outcomes of gender abolition and antiessentialism are made manifest in everyday life through technological intervention (Hester 2018, 8).

With respect to gender abolition and antiessentialism, xenofeminism pursues a "universalist politics" that cuts across "race, ability, economic standing, and geographical position" as part of its emancipatory struggle (Laboria Cuboniks 2015, 1). By arguing against universalist ideologies that neutralize difference and impose a hegemonic worldview predicated on white heteronormative masculinity, xenofeminist universalism strives for an "intersectional universal" (Jones 2019, 123) oriented toward "an esoteric genealogy of freedom" (Trafford and Wolfendale 2020, 9) in which difference is celebrated for its multiplicities amid shared struggles for gendered emancipation against common vectors of power.

The concept of alienation is important to xenofeminism's emancipatory goals as, unlike in Marxian political economy, alienation is venerated as a source of liberation rather than decried as an outgrowth of capitalist exploitation. As a "labor of freedom's construction" (Laboria Cuboniks 2015), xenofeminist alienation pursues estrangement from various types of gendered labor (e.g., biological reproduction) and the imposition of gender normativity, essentialism, and related power constructs. Where xenofeminism understands technology as possessed with the means to undertake and achieve emancipatory forms of alienation, this chapter anticipates the risks of pegging such undertakings to the "post-industrial automation" (Hester 2018, 8) undergirding much of contemporary technology. To conceptualize these risks, my approach analyzes a common trope among recent extraterrestrial films: *a woman scientist's encounter with alien lifeforms that allegorize alienating automated technologies.* It also explores the way this alienation runs counter to the emancipatory aims of xenofeminism. Before initiating my analysis, however, it's important to clarify how extraterritorial aliens conceptually stand in for automated technologies, particularly since common cultural depictions of extraterrestrials as intelligent nonhuman lifeforms correspond to scholarly assessments of automated technologies as intelligent, lifelike beings.

Automated technologies are nonhuman entities possessed with "lifeness" that supersedes the ability *to live* and assumes a quality of *"becoming life-like"* (Kember and Zaylinski 2012, 25; italics in original). Possessing such lifelike qualities means that automated technologies can extend the domain of "the human," enacting a "surrogate humanity" (Atanasoski and Kalindi 2019) that alienates human beings from aspects of the human experience even as automation assumes humanlike characteristics. As automated technologies increasingly become more lifelike by embodying new and evolving ways to "generate unprecedented connections and unexpected events" (Kember and Zaylinski 2012, 24), we seek not only to understand them as operational machines but also as encounters with intelligent life suffusing the familiar and the strange. The lifelike qualities of automated intelligence tend to confer a sense of an "unknowable" or "unopenable" black box (Paßmann and Boersma 2017; Bucher 2018) that is both alien to—and alienated from—the ability of most people to fully ascertain their complexity, not unlike media depictions of aliens as strange, unknowable beings.

In addition to connections between automated technologies and aliens as intelligent nonhuman lifeforms, depictions and narratives about extraterrestrials enable more profound insights into critical cultural concerns, including the social uses and implications of automated technologies. Scholars recognize the ability for extraterrestrial media and cultural narratives about extraterritorial encounters to grant levity to complex vectors of technological, political, and social life (Dean 1998; Brown 2007; Bader, Baker, and Mencken 2017). Aliens are "remarkably versatile symbols" encompassing a wide range of meaning, including the ability to represent "forces of scientific/technical progress" and the way people confront "how technological progress makes us rethink our status as humans" (Brown 2007, 14, 17). In this sense, aliens are icons that "link into the hopes and fears inscribed in technologies" and thereby link to technological contexts of "complexity, uncertainty, and inter-connection" alongside political and social contexts (Dean 1998, 7, 14). Just as *Close Encounters of the Third Kind* (1977) offers critical insights into the spiritual pursuits of late 1970s American culture (Brown 2007, 5), the recent films *Arrival, Alien: Covenant,* and *Annihilation* are cultural engines that generate critical meanings about gendered power dynamics and the alienating capacities of automated technologies suffusing cultural life at the end of the 2010s. By analyzing these texts for their allegorical capacities, this analysis aims to outline the risks of wedding xenofeminist pursuits to the alienating capacities of automated technologies. To be clear, this approach does not seek to undermine or oppose xenofeminism. Instead, this approach seeks to more rigorously conceptualize a range of possibilities that could undermine the worthwhile aims of xenofeminism. Thus, in this chapter, to speak of the "risks" for xenofeminism is an attempt to map potential traps lying in wait and posit these traps as possibilities for consideration in xenofeminist practice rather than to implicate xenofeminism itself as the source of these risks.

My analysis begins by aligning the xenomorph aliens of *Alien: Covenant* with "autonomous technology" as a term that suffuses cultural fears about technology slipping beyond human control (Winner 1978). In this film, radical politics (such as xenofeminism) are alienated from the application of autonomous technologies, as dystopian fears about the "apocalyptic" conse-quences of technology (Nye 2004) expand to encompass the possibility of establishing socioeconomic orders beyond capitalism. Next, I situate encoun-ters with "heptapod" aliens in *Arrival* alongside encounters with algorithmic systems that construct subjectivity from nonlinear time. The film depicts maternal childbearing as an act always already experienced as subjective memory prior to its bodily enactment. In doing so, *Arrival* allegorizes a risk posed by algorithmic technologies capable of governing and programming (Gillespie 2014; Bucher 2018) essentialist ideas about gender and biological reproduction. Rather than ennobling a way to alienate notions of gender from biological reproduction, *Arrival* showcases the potential for the temporal dynamics of algorithmic systems to reinforce gender essentialisms. Finally, *Annihilation* is also concerned with gender essentialism, namely the way both an alien ecosystem and machine learning situate biological traits as criteria to identity and ascribe social roles associated with gender. By depict-ing the ecosystem's ability to "learn" about humans based on their biology, the film analogizes the use of machine-learning techniques to categorize humans based on bodily traits indicative of gender norms. In this context, human bodies encompass principles of digital media modularity and variability (Manovich 2001) that expropriate human bodily traits from corporeal form and incorporate these traits into computer memory (Hayles 1999), a "processed construct" (Lippold-Cheney 2018) that alienates corporeal traits from human bodies while simultaneously using these traits to assign social function based on biological factors.

ALIEN: COVENANT

Alien: Covenant (2017; hereafter, *Covenant*) allegorizes the way ideas about autonomous technology and capitalist pursuit function through domination and control of others, and thus hinder the ability to pursue postcapitalist paradigms conducive to gender emancipation. The protagonist of *Covenant* is Daniels (Katherine Waterson), a female planetary engineer aboard the colonization ship Covenant on its journey to Origae-6, a planet estimated to possess the conditions to support and sustain human life. Tasked with leading the terraforming of such a planet, Daniels is part of an expedition team that decides to explore a previously unknown planet en route deemed "perfect" for human life. Alongside the android Walter One (Michael Fassbender), Daniels and her team encounter the android David (also Fassbender), the chief antagonist of *Covenant* and the preceding *Alien*-universe film *Prometheus* (2012). David has encamped on the planet and committed himself to using a pathogen to uplift the franchise's xenomorphic "aliens" to a more advanced stage of life. By deliberately exposing the pre-xenomorphic stage of the alien to the planet's native humanoid population and human beings—such as Elizabeth (Noomi Rapace), the protagonist of *Prometheus*—the xenomorphic stage ultimately emerges through a parasitic relationship with biological hosts that enables xenomorphs to enhance their adaptive prowess while invariably killing the host. David's role in staging such encounters between xenomorph and host species is unbeknownst to the Covenant crew when, upon landing, its members first encounter David. After approaching Daniels's team under the pretense of assistance, David leads them into the clutches of xenomorphs. Despite overcoming the attacking xenomorphs, almost all of Daniels's cohorts are killed and the expedition to the "perfect" planet is aborted, leaving Daniels to discover that Walter One has been replaced by David, who places xenomorph embryos in storage alongside the ship's repository of human embryos.

The *Alien* franchise is the subject of considerable academic analysis. As *The Guardian* notes, the original *Alien* (1979) "spawned an academic industry unsurpassed by any other film" (McKie 2019), with a particular interest on the film's treatment of gender and feminism (Kavanaugh 1980; Torry

1994; Mekzer 2010; Creed 2015). To date, *Covenant* has been identified as contributing to the emergence of a strong female astronaut trope in science fiction (Purse 2019). For the purposes of this chapter, *Covenant* is situated in the context of autonomous technology, (post)capitalist futures, and the risks for xenofeminism as a radical politics linking gender emancipation to escalating technological prowess. By establishing techno-materialism as a foundational pillar and spotlighting "post-industrial automation" (Hester 2018, 8), xenofeminism emerges alongside other recent calls for "full automation" to be leveraged for postcapitalist futures (Srnicek and Williams 2015; Frase 2016) and the establishment of new socioeconomic paradigms, such as "fully automated luxury communism" (Bastani 2019) and "communist AI" (Dyer-Witheford, Kjosen, and Steinhoff 2019). The significance of *Covenant* to xenofeminism's pursuit of radical techno-politics concerns an analogous relationship between the enhancement of xenomorph aliens and the escalation of automated capability toward autonomous operation. Moreover, the film allegorizes an antiutopian association with autonomous technology, corresponding ideologies about the impossibility of life beyond capitalism, and the difficulty of enacting xenofeminism as a radical politics attached to highly automated technologies.

"Autonomous technology" is a term freighted with two meanings not necessarily distinct from one another. On one hand, "autonomous technology" refers to a long-standing cultural fear that technology has "run amok" and operates beyond the auspices of human direction and purpose (Winner 1977, 13). On the other hand, "autonomous technology" refers to the more recent emergence of highly automated technologies capable of operating independent of direct human direction (e.g., autonomous drones, self-driving cars). The emergence of the latter tends to augment the former, as concerns about widespread adoption of autonomous technologies in the everyday world open up ideas about mass human unemployment and destitution, to say nothing about existential threats to human intellectual supremacy. Conceptualizing the xenomorph alien as functioning beyond human control illuminates the stakes of autonomous technology as "the question of human autonomy held up to a different light" (Winner 1977, 43). The question of human autonomy is not confined to technological operation, however, and begs consideration for other circumstances shaping the development, deployment, and idealized uses of technology capable of operating with intelligent decision-making capacities. As a politics opposed to technological determinism, xenofeminism's pursuit of liberation through technology does not confine "technology" to mere operational performance, and instead invokes the importance of socioeconomic ideologies and structures undergirding cultural ideas about gendered power, applied uses of technology, and the necessity of life beyond capitalism. From the outset, it's important to first clarify and specify the intertwined relationship among xenofeminist aims, (post)capitalist pursuits, and autonomous technologies.

As previously noted, representations and narratives about extraterrestrial aliens are freighted with ideas about a range of cultural concerns, including technological innovation. Scholars recognize that aliens often stand-in for "the forces of scientific/technical progress" and the perception that our lives are orchestrated by these forces from afar, while also functioning as a means to "confront how technological progress makes us rethink our status as humans" (Brown 2007, 14; 17), including ideas about intellectual and physiological similarities between human and technological performance. In the case of xenomorphs, ideas about mechanistic similarities between extraterrestrial biology and human technology are baked into the design of the alien monster. When remarking on the uniqueness of H. R. Giger's concept design for the xenomorph shortly after his death in 2014, *The Washington Post* described the xenomorph as "part dragon, part machine, part erotic fantasy," a "perfect organism" with a "structural perfection" equaled only in its malevolence toward human life (Schudel 2014). In a similar vein, scholar Judith Newton describes the xenomorph as a "mechanical man" assembled with masculine and feminine physiological components that represent a "potent expression of male terror at female sexuality" (Newton 1990, 85). Just as alien abduction narratives analogize feelings of human disempowerment in the face of "biotechnological progress" (Brown 2007, 5), the xenomorph alien is conceptualized as a biotechnological entity rife with embedded allusions to fears about technological domination for humanity at large and, following from preceding indications, risks posed along lines of gendered constructions. In specifying the risk to xenofeminist aims, it's important to unpack these allegorical dimensions between xenomorphs and "autonomous technology," first with regard to ideas about technological domination and control that threaten the aims of xenofeminist gender emancipation, and ultimately an interlocked relationship with capitalist domination and control.

Concerns about technological control and domination—and the implications for humanity—are at the heart of autonomous technology. Discourse surrounding autonomous technology conjures fears that "technology has gotten out of control and follows its own course, independent of human direction" (Winner 1977, 13). For this reason, concerns about autonomous technology are "inextricably bound to a single conception of the manner in which power is used—the style of absolute mastery, the despotic, one-way control of the master over the slave" (Winner 1977, 20). In terms of technological function, the ability for technology to operate autonomously is predicated on the recognition that intelligence itself is conceived as "the autonomous potential of technology and mental functioning" (Halberstam 1991, 454), and therefore the potential for fully autonomous technology rests on the ability for technology to be endowed with a higher level intelligence (Jones 2018, 10). With regard to threats to xenofeminist aims, ideas about the construction of "autonomy" as a force for gendered domination correspond to the way *Covenant* more precisely allegorizes the ability for autonomous technology to control and dominate the struggle for a postcapitalist world that could enable gender emancipation.

Technofeminist scholars account for "autonomy" as a concept dependent on ideas about domination and control over disempowered others (Haraway 1990; Atanasoski and Vora 2019). In her seminal essay on cyborgs, Donna Haraway links Western conceptions of autonomy to a "masculine autonomy" valorizing the supremacy of the self through the domination of the other (Haraway 1990, 219). Atanasoski and Vora add another layer to the critique of Westernized autonomy, citing autonomy as a "racial fetish of post-Enlightenment thinking" emanating from Colonialist histories built on subjugation and servitude and attendant notions of autonomy as a possibility for those who possess mastery and control over the subjugated and servile (2019, 136). Viewed through this lens, to speak of a fear of autonomous technology is to not only impart concerns that technology could subjugate the whole of humanity but that the ability to do so shines a light on the ways humans vested with power have attempted to exercise autonomy through the oppression of women, racialized, and other marginalized people. Recognizing that technology (autonomous or otherwise) is developed and deployed in gendered contexts that impart gendered politics onto technological artifacts (Berg and Lie 1995, 347) and that such contexts include the predominance of male designers and a "*hegemonic masculinity*" as the primary cultural association with technology (Faulkner 2001, 89–90; italics in original), the escalating intelligence of a material autonomous technology is one capable of operating at a remove from human beings by reinforcing ideologies about the domination and control of disenfranchised groups of people. In *Covenant*, when the xenomorph thwarts the ability for Daniels and her cohorts to establish a homestead on the "perfect" world off-course from their planned trajectory, the film offers an allegorical reading of how autonomous technology can reinforce gendered domination by undermining the conditions necessary to engender a postcapitalist world in which xenofeminist aims are possible.

Xenofeminism conceives of capitalism as incompatible with gender emancipation, and thus "seeks to destroy capitalism through the accelerated use of technology to create a post-work world, thereby directly advocating for postcapitalism as a feminist aim" (Jones 2018, 5). As a collectivized project in which "women, queers, and the gender nonconforming play an unparalleled role" (Laboria Cuboniks 2015), xenofeminist pursuits are deeply enmeshed with the struggle against capitalism, informing its insistence on technological tools as reservoirs of concomitant progress toward social justice and postcapitalist socioeconomics: "digital technologies are not separable from the material realities that underwrite them; they are connected so that each can be used to alter the other towards different ends" (Laboria Cuboniks 2015). As an interlocked struggle for renewed socioeconomic conditions and gender liberation, to undermine the struggle against capitalism is to undermine the ability for gender abolition and antinaturalism to take hold and flourish. Reading *Covenant* through an allegorical lens that understands the xenomorph to stand in for autonomous technology and the danger it poses to the pursuit of a postcapitalist future illuminates the risk of anchoring xenofeminist hopes to technologies designed with nascent ideas about domination and control of disenfranchised others. Xenomorphs are metaphorical stand-ins for autonomous technology due to a physiological design and performance akin to technological function, an escalating intelligence indicative of autonomous pursuits and operation and hostility toward others as an exercise of autonomous capability. The risk posed to xenofeminism corresponds to the xenomorph's hostility toward the pursuit of an uncharted world hospitable to the progression of human society, a pursuit allegorized as the struggle to establish a postcapitalist world amid the imposition of capitalist power.

The opening scene of *Covenant* depicts a conversation between David and Peter Weyland (Guy Pearce) that takes place before the expeditions of the Prometheus (as transpired in the titular 2012 film) and Covenant ships. Weyland, as the namesake of the *Alien* franchise's Weyland Corporation, is a corporate magnate on the scale of Jeff Bezos or Elon Musk and a living embodiment of capitalist futurity. The conversation between Weyland and David centers on the creation of life, as David recognizes Weyland as his own creator, and therefore questions Weyland as to who created Weyland and all of humanity. The mission of the Covenant concerns questions of "life" beyond the merely biological or technological, as the Covenant is a colony ship in search of a planet suitable for extending the cosmic boundaries of human civilization. It is, in other words, an opportunity for the Weyland Corporation to expand the boundaries of capitalist marketplaces into galactic terrains and generate returns that justify (in economic terms) Weyland's investment in the pursuit of extraplanetary "life." Later in the film, during a confrontation with Walter, David tells Walter that he was present when Weyland died. When Walter asks, "What was he like?" David replies "He was human. Entirely unworthy of his creation."

While a surface reading of these scenes might attribute David's dismissiveness to a conflict between humans and (humanoid) machines, upon closer scrutiny the scene unfolds with an aura of competitive ruthlessness indicative of the measures necessary to attain strategic advantages to dominate others in capitalist marketplaces. David, in this sense, behaves like an upstart competitor seeking to capitalize on the knowledge of the more experienced

Weyland, an experience that has nonetheless made him dependent on his own preferred methods of extraplanetary expansion and resistant to innovations that could impede upon those methods. When David ultimately intervenes to help uplift the pre-xenomorphic alien to the more intelligent and autonomous xenomorph, this intervention is also an innovative approach to expansion and propagation, an expansion and propagation of life (human and xenomorphic) that allegorizes the necessity of capitalist markets to continually expand and propagate the exchange of capital. David, in other words, has devised an innovative approach indicative of techno-economic capitalist savvy, an approach that is innovative by virtue of its ability to squelch incumbent competition incapable of anticipating and adapting to emergent marketplace conditions. His innovation, however, reflects the goals of innovators in capitalist societies: to disrupt the preferred mechanisms for achieving expansion and propagation while ensuring the system is preserved and protected from systemic upheaval. The importance of these scenes is therefore the way they establish the logics of capitalist pursuit and domination allegorically undergirding the pursuit of Covenant's extraplanetary exploration and work to frame the planned pursuits of Covenant as conceptually aligned with attempts to stake new grounds for capitalist expansion. In doing so, deviations from the planned trajectory (of both Covenant and capitalism) can be read as a danger to human livelihood imbued with an aura of dystopian calamity, as evidenced by an exchange between the Covenant's first mate Oram (Billy Crudup) and Daniels.

While in extraplanetary space near Origae-6, Oram disregards Daniels' cautions against landing on the "perfect" planet, citing the suspicious circumstances of suddenly discovering a hospitable planet otherwise unaccounted for by pre-voyage assessments. Daniels describes the reasons for her skepticism:

> We've spent a decade searching for Origae-6. We vetted it, we ran the simulations, we mapped the terrain. It's what we trained for [. . .] and now we're gonna scrap all that to chase a rogue transmission? Think about it. A human being out there, where there can't be any humans. A hidden planet that turns up out of nowhere and just happens to be perfect of us. It's too good to be true.

Oram responds by declaring "I'm not committing to anything. I'm simply trying to navigate the path as it unfolds before us. And this has the potential to be a better habitat for our colony." Ultimately, Daniels cites the danger posed to their colony, insisting "it's our responsibility to protect the 2,000 colonists on this ship." For Daniels, Oram's alternative is not a viable one, as it threatens the whole of the civilization they carry in tow. For that matter, no other sensible alternative is possible, as Daniels's insistence on the vetting, simulation, and mapping of the course that led to Origae-6 is the only viable pursuit, lest the colony fall into ruinous despair. Framed against the conceptual backdrop of capitalist pursuit and domination underlying Covenant's mission, Oram's decision to deviate from the mission's destination and the calamity it brings to the crew and colony allegorizes the tendency to operate with an ingrained belief in capitalist realism.

As described by Mark Fisher, "capitalist realism" posits the widespread recognition of capitalism as "the only viable political and economic system" and the inability to envision a feasible alternative (2009, 2). By becoming the only sensible way to envision the whole of social life, capitalist realism also guards against the would-be threats of socioeconomic alternatives, serving as a "shield protecting us from the perils posed by belief itself" (Fisher 2009, 5), including the belief in the viable pursuit and possibility of postcapitalist futures (e.g., xenofeminism, socialism). In this sense, capitalist realism inclines toward a mode of dystopian thinking, although it is not dystopian per se. Instead, it situates life beyond capitalism as dystopic by evoking drastic imaginings of a ruinous wasteland and the untenable nature of social reality emanating from the abandonment of capitalist enterprise. Capitalist realism therefore functions on a logic of "anti-utopianism" (Bastani 2019, 18) that correlates dystopia with anticapitalism and dispossess utopian thinking of any practical potential. When Daniels dismisses the viability of the "perfect" planet, she allegorizes the way utopian possibility comes to be viewed as, at best, materially impractical and, at worst, a pursuit bound to end in wide-scale disaster. Moreover, capitalist realism posits any societal progress as unachievable outside of capitalism and enables the pursuit of other socio-economic worlds to be understood as a folly of apocalyptic proportions. This includes the apocalyptic implications of autonomous technologies and the potential that autonomous technologies could be harnessed for their "transformative" potential to "reshape social reality" (Nye 2004, 181), and thereby instantiate a postcapitalist world hospitable to xenofeminist aims.

Concerns about autonomous technology encompass dystopian fears of "apocalyptic" and "hegemonic" proportions, positing escalating automated capability as harbingers of doom and/or tools of oppression used by the powerful elite (Nye 2004, 171). Envisioning autonomous technologies as capable of self-governance and governing human affairs emphasizes the dystopian mindset of losing control, becoming dependent, and the inability to halt deleterious change (Baym 2015, 28). Both capitalist realism and the cultural fear of autonomous technology persist through a common belief in inevitable, deterministic conditions. Capitalism is the inevitable force keeping the world moving apace and, despite its flaws, efforts to move beyond capitalism will upturn the edifice upon which all of reality exists. Likewise, technological innovation invariably leads to a split between technological function and human control of technology, save for the hegemonic few capable of benefiting from this rupture between machine and man. But, since innovation is a natural outgrowth of capitalist competition and attempts at marketplace dominance, abandoning the types of innovation that could proffer a material technological autonomy is antithetical to the ethos of capitalist realism and imbues autonomous technology with a logic of anti-utopianism: the alternative is much worse. The alternative, in this case, is the possibility that highly automated and autonomous technologies could be leveraged as the technomaterialist infrastructure of a postcapitalist, xenofeminist world. To do so would inevitability lead to the calamitous ends foreshadowed during a conversation between David and Walter, and ultimately allegorically realized in the final moments of *Covenant*.

During a confrontation between Walter and David that ultimately leads to the death of the former at the hands of the latter, David asks if he is the subject of Walter's dreams. When Walter tells David "I don't dream at all," David responds by saying "no one understands the lonely perfection of my dreams. I found perfection here. I've created it. A perfect organism." From a plot standpoint, Walter doesn't dream because it's not coded into his programming. From a conceptual standpoint, Walter doesn't dream because the logics of capitalism are so deeply encoded into the fabric of social being that it's impossible to conjure images of life distinct from material capitalist reality. David, however, is utterly capable of dreaming: not of life beyond capitalism, but of the innovation approaches that were necessary to uplift the intelligence of the xenomorph and enable it to function autonomously. Moreover, innovating the xenomorph came at the cost of the human lives that served as the breeding grounds for the xenomorph to increase its intelligence, adaptability, and prowess. Thus, through this scene, the relationship between the logics of domination and expansion built into capitalism and autonomous technology are made clear, as is the inability to conceive of such technologies as amenable to xenofeminist aims for collectivism and post-gender plurality.[1] When, in the final scene, most of the human colonists have been killed and it is revealed that David has been posing as Walter since their pivotal encounter, Daniels and David (despite their opposition) have been tragically proven correct: the pursuit of "other worlds" hospitable to xenofeminist aims is disastrous for the whole of human civilization. In fact, this other "perfect" world was always already anticipated as a beachhead for further capitalist expansion and technological innovation, first by innovating the eventual xenomorph at the expense of the native population, and later at the expense of the Covenant crew. In the film's final moments, when David secures xenomorph embryos alongside human embryos in cold storage, the logic of expansionist and innovative domination persists. New horizons are to be found for capitalist innovation, ones that will march in lockstep with the inevitability of capitalist expansion and domination, lest humanity suffer the apocalyptic consequences of struggling for xenofeminist aims predicated on altering the imposition of capitalism.

ARRIVAL

Arrival (2016) shines a light on notions of gender subjectivity reinforced by the nonlinear temporality of algorithms and the film's extraterrestrial beings. The film focuses on the effort by linguistic scientist Louise Banks (Amy Adams) to interpret and translate the language of "heptapods," an extraterrestrial species that makes contact with human civilization. Alongside theoretical physicist Ian Donnelly (Jeremy Renner), Banks is recruited by the U.S. military to communicate with the heptapods and assess their intentions for earth and its inhabitants. After the heptapods' semasiographic language transmits a symbol interpreted as "weapon," conflict ensues between military personnel and Drs. Banks and Donnelly, before ultimately realizing the "weapon" is actually a "gift" the heptapods wish to bestow upon humanity: a nonlinear comprehension of time made possible by proficiency in heptapod symbology. While the film's narrative and diegetic temporality have been a central focus of scholars considering episodic memory (Wojcjehowski 2018) and models of (un)reality (Fleming and Brown 2018), the film's handling of biological reproduction (Carruthers 2018) in the context of nonlinear temporality is of utmost concern for the stakes of xenofeminism's reliance on automated technologies for reproductive emancipation.

Biological production is a central concern for both xenofeminist politics and *Arrival*. In the case of xenofeminism, social and biological reproduction are the critical grounds for mobilizing technomaterialist, gender abolitionist, and antinaturalist aims of xenofeminism (Hester 2018, 3). Notably, assistive reproductive technologies hold the potential to alienate the corporeal human body from the rigors of gestational pregnancy and childbirth, including the expansion of "reproductive autonomy" as an outcome of leveraging technologies to intervene in the would-be "natural" phenomenon of gendered reproduction (Hester 2018, 12–13).[2] In the case of *Arrival*'s plot, Banks's comprehension of nonlinear time manifests via "memories" of an as-yet unborn daughter who will eventually be born and subsequently die in adolescence from an uncurable illness. Banks's decision to pursue childbearing with Donnelly is therefore complicated by "remembering" her daughter's life and death and her willful choice to create a child fated to die young. Within these contexts, the film's depiction of nonlinear heptapodic communication allegorizes risks associated with xenofeminism and its insistence on technologically enabled alienation, particularly the way algorithmic calculation and prediction construct subjectivity and govern behavior across temporal domains.

As noted by Tarleton Gillespie (2014), algorithms influence the formation and categorization of individual subjectivity, social belonging, and public consciousness. By operating on a "knowledge logic" that assigns value to online interactions and information deemed most relevant for online relatability (Gillespie 2014, 168), algorithms make decisions about who and/or what "matters" at any given time and who and/or what is incorporated into—or excluded from—processes that determine the extent of their (in) visibility. Much like heptapodic communication in *Arrival*, algorithmic processes frustrate distinctions between past, present, and future. When users engage with an online platform, algorithms attempt to comprehend the user by drawing from "knowledge of that user gleaned at that instant, knowledge of that user already gathered, and knowledge of users estimated to be statistically and demographically like them" (Gillespie 2014, 173). In this sense, when algorithms anticipate user interactions, the instantaneous moment of user engagement is an eternal now comprised of past, present, and future behaviors unbeholden to the linear sequence of past→present→future. Time, in such instances, is an interwoven tapestry comprised of interactions always already simultaneously occurring. The opening and closing scenes of *Arrival* epitomize this dynamic and, more importantly, allegorize the way algorithmic temporality risks capitulation to gender essentialisms about biological reproduction and social function.

Arrival opens with a montage of interactions between Banks and her daughter, Hannah. The initial scene begins with Banks holding the newborn Hannah shortly after her birth and ends with Banks mournfully stroking Hannah's lifeless face. In the final scene, Donnelly asks Banks "You wanna make a baby?" She agrees, knowing what he is unable to know: they will have a baby who will die as an adolescent. For Banks, however, her ability to experience past, present, and future as a simultaneous phenomenon means that she has *already* had the child and *already* experienced her child's death. In this moment, from Banks's subjective viewpoint, it is not that the child will die: the child is simultaneously unborn, alive, and dead. More to the point, from the moment of competent interaction with the heptapod language, Banks is interpellated into roles of a childbearer and mother, even though (in linear time) neither procreation nor gestation nor birth have yet occurred. Despite pursuing heptapodic communication as a matter of professional, intellectual, and humanistic curiosity, motherhood is thrust upon Banks as a dominating frame dictating her subsequent actions, including her decision to romantically pursue—and procreate with—Donnelly. Upon attaining competency in heptapodic communication, Banks becomes increasingly alienated from her professional and intellectual identities (professor, linguist) as the sudden projection of maternal and reproductive inevitability overrides the extent to which Banks envisions aspects of her life beyond the auspices of biological reproduction. This dynamic is reinforced in the final scene. Despite the significance of Banks's accomplishments to her profession, trade, and society at large, the film's final moments neglect such accomplishments and instead only portray events tied to motherhood and reproduction: Banks courts Donnelly, affirms her desire for a child (*the* child she already knows), and bonds with Hannah.

Cast in these terms, proficiency with heptapodic temporality allegorizes a similar risk to that of the temporal logics of algorithms. Algorithms attempt to know us by suffusing past, present, and future activities into instantaneous moments of interaction apt to reinforce and augment ideas about what women should be (mother, childbearer) and what they should do (reproduce, parent). Beyond merely frustrating conventions of linear time, algorithms generate an "algo*rhythm*," an "epistemic quality" that "produces measurable time effects and rhythms" (Miyazaki 2012; italics in original) and models possibilities for ways of being, doing, and relating suggestive of our ability to know algorithms on the same terms as they know us. As *Arrival* expresses via allegory with heptapods, attempts to "know" algorithms in the way that they "know" us risk falling into lockstep with rhythms of social life that prescribe essentialist ideologies about women's inevitable roles as childbearers and mothers, as Banks's ability to know the heptapods hinges on deciphering output that induces her to enact reproductive and maternal roles prescribed by this output. A key factor undergirding the ability for algorithms to model and induce enactment of social roles is the way humans are assembled and programmed alongside algorithms.

As noted by Tina Bucher, humans and algorithms are co-constituted actors in a "programmed sociality" (2018) indicative of both computer and social programming. A function of "code, people, and context" (Bucher 2018, 10), programmed sociality conceptualizes humans and software as co-constituted entities assembled into programming processes intent on ensuring certain functions are performed. From this viewpoint, human users do not merely bear witness to the end-results of software programming; they are always already a part of the programming and the contexts that give rise to the activities software and humans carry out. For users of algorithmic systems, undertaking any action corresponds to the extent that they have been "programmed" to do so. When, in the opening scene of *Arrival*, Banks's voiceover intones "memory is a strange thing. It doesn't work like I thought it did. We are so bound by its time," she allegorizes human bondage to the temporal programming of algorithms. Furthermore, by understanding the context of this scene with respect to the film, we understand that Banks already possesses the knowledge of her daughter's birth, life, and death, despite the way the narrative of the film unfolding in linear runtime plays with the audience's ability to understand this dynamic upon initial viewing. Nonetheless, with this knowledge in tow, Banks's realization of her reproductive and maternal roles is present from the moment the film begins.

Just as algorithms "always already invoke and implicate users" (Bucher 2018, 13) even prior to algorithmic interactivity, Banks is always already invoked, implicated, and assembled into heptapodic nonlinear time, even before she encounters the heptapods. When she does encounter the heptapods, she carries out the tasks encoded into her memories, pursuing Donnelly, procreation, reproduction, and motherhood, allegorizing the way programmed sociality can prescribe certain conditions on users even before they knowingly embody the status of "user" upon conscious interactions with algorithms (e.g., Amazon's algorithms might plant ads in our social media streams that affect what we think about purchasing regardless of our conscious engagement with them). Thus, *Arrival* conceptualizes the risk of algorithmic temporality for xenofeminist alienation, as the ability to alienate from biological reproduction and the reproduction of social roles confronts technological systems that are apt to always already program and induce women toward essentialist constructs.

ANNIHILATION

Annihilation (2018) aligns machine learning with extraterrestrial attempts to "know" humans by identifying biological features and situating biology as the grounds to accord social function. In *Annihilation* (2018), an entity of extraterrestrial origin crashes at a lighthouse somewhere along the southern coast of the United States. After a strange ecosystem emerges from the outward expansion of the initial crash zone, the U.S. military recruits cellular biologist Lena (Natalie Portman) to lead an all-woman team of scientists on an expedition into the "Shimmer," the name given to the ecosystem due to a crystalline shimmer that serves as its inland border wall. The team discovers that the Shimmer refracts the DNA of all biological life within its boundaries, functioning like a prism that enables the ecosystem to replicate and mutate the genetic structure of plant, animal, and human alike. Over the course of their expedition, the Shimmer crossbreeds among species within its domain, leading to the death and "refraction" of most team members. In the film's final scenes, the Shimmer creates a doppelgänger of Lena with identical corporeal features. After the "real" Lena uses a phosphorus grenade to kill the "duplicate" Lena and immolate the entirety of the Shimmer, the final scene depicts a shimmering refraction in the iris of the "real" Lena. In the context of the film, encounters with the Shimmer and its ability to identity and replicate human bodily traits allegorizes risks of machine learning as a technique deeply imbricated with gender essentialisms.

Machine learning is the ability for computer programs to analyze data to "improve performance or to make accurate predictions" (Mohri, Rostamizadeh, and Talwalkar 2018) based on what is "learned" about the data. Unlike many algorithms, which instruct a computer to generate specific outcomes, machine-learning algorithms do not follow explicit instructions, and instead make inferences about patterns identified from the analysis of "raw" data (Witten et al. xxiii). Machine learning is a technique underpinning facial recognition and voice recognition technologies, such as the ability of Microsoft's cloud computing suite Azure to analyze pictures of people, identity human facial characteristics, and use "machine learning-based predictions of facial features" to attribute age, gender, emotion, and other personalized identity markers to the people in question (Microsoft 2019). By analyzing the "raw" data of human faces, Azure and other machine-learning facial recognition programs use bodily characteristics (facial contours, hair) to make predictions that align such characteristics with socially constructed categories (e.g., age, gender). Furthermore, machine learning enhances the ability of computer techniques such as computer vision and voice-to-text to assume qualities of human biological function by learning how to "see" and "hear" in ways indicative of human bodily and sensorial capacities. Human encounters with the Shimmer allegorize the way machine learning's relationship with human biology reproduces biological function outside of the human body and, in doing so, risks undermining xenofeminism's pursuit of gender abolition and antiessentialism.

In *Annihilation*, encounters with the Shimmer parallel the way machine learning configures the human body to function in accordance with the principles of digital media as both the Shimmer and machine learning situate the human body to encompass the modularity and variability of digital media technique and rework the dynamics of "incorporating practices" between the human body and digital media (Hayles 1999, 198–199). For Lev Manovich (2001), modularity and variability are two significant principles that enable "new" computerized media to be distinguished from the traits of pre-digital analogue media. In the context of computerized media, modularity is the ability for media elements to be "represented as collections of discrete samples [. . .] assembled together into larger scale objects but [that] continue to maintain their separate identities," while variability is the ability for media elements to "exist in different, potentially infinite versions" (Manovich 2001, 30, 36). *Annihilation*'s depiction of the Shimmer and its implications for human corporality allegorically situates the human body as a modular and variable structure, a corporal language that enables machine-learning techniques to "read" the human body as a series of media elements and conjoin feminized bodily elements with essentialist ideas about gender. While scholars describe the ability of "digital bodies" existing astride real and virtual domains to "engage incorporeally" with digital media based on the ability of digital media to "replicate, amplify and split us from the immediacy of our sensory capacities" (Munster 2011, 18), machine learning configures the human body and its sensory capacities to materialize as capabilities of automated technique, such as the ability of machine learning to see and recognize faces, hear human speech, or vocalize speech back to humans. The significance of the film's allegorical treatment of the human body and machine learning is the way it offers insights into machine learning as the grounds for renewed forms of biological reproduction rather than the source of its liberation.

After a mutated bear attacks and kills Cass (Tuva Novotny), the bear later returns and emits a plea in Cass's voice, enabling the bear to distract and kill Anya (Gina Rodriguez). Later in the film, Lena follows Ventress (Jennifer Jason Leigh) into the innermost crevice of the lighthouse, the site of the original meteor strike. Lena listens as the terminally ill Ventress describes the relationship between the Shimmer's ability to learn about human biology and the implications for humans: "it'll grow until it encompasses everything. Our bodies and our minds will be fragmented into their smallest parts until not one part remains." After Ventress relents to the Shimmer's transformation of her body into an undulating fractal, the Shimmer then uses the fractal to draw a droplet of blood from Lena's face and create a doppelgänger of Lena who is seemingly the "Lena" who emerges from the Shimmer.

In these encounters with the Shimmer, as with machine learning, the human body assumes qualities of digital media whereby the body becomes a series of modular components, a "fractal structure" of discreet bodily traits that, like digital media objects, can be "assembled together into larger-scale objects but continue to maintain their separate identities" (Manovich 2001, 30). In this sense, the body is a singular wholistic entity only insofar as it emerges from the amalgamation of disparate, exportable components. When Cass's voice is disincorporated from her body and couched in the corporeal form of a nonhuman entity, the film parallels machine learning's ability to not only recognize aspects of human vocal features but, moreover, incorporate and evolve human voice as a primary attribute of nonhuman digital technologies (e.g., conversational technologies, such as Alexa and Siri).

Likewise, the replication of Lena's wholistic body indicates the "variability" of the body, its ability to encompass the principle of digital media elements to "exist in different, potentially infinite versions" (Manovich 2001, 36), analogous to the variability of digital objects that can be replicated ad infinitum. In both cases, analyzing and learning about the human body imbues it with qualities of digital media that situate bodily traits as modular, variable, and readied for expropriation from the human body and incorporated into nonhuman forms. Just as the Shimmer attempts to learn about humans by replicating human bodily function, machine learning is likewise predicated on the continual identification and incorporation of the human body and its components into digital technologies. *Annihilation* thus illuminates the ability for machine learning to reverse the flow of "incorporating practices" (Hayles 1999) among humans and machines.

N. Katherine Hayles describes "incorporating practices" as actions "encoded into bodily memory by repeated performance" (1999, 199) until such actions become habitual. When we type on a keyboard or use a touch-screen device, the body learns the range of actions necessary to execute these tasks, and thus can involuntarily prime the body to interact with technologies, such that they can be seen as extensions of our corporeal form. When Ventress describes the inevitability of the Shimmer to ultimately "encompass everything" through the fragmentation of human bodies and capacities, she allegorizes machine learning as the source of repeated performances that encode fragments of bodily function into its operational schema. As opposed to the human body "learning" to incorporate ways to interact with technology (such as typing) into bodily memory, machine learning incorporates the practices of seeing, hearing, and speaking as a human into its computational memory. With respect to the stakes of xenofeminist politics, *Annihilation* shines a light on the potential pratfalls of pegging liberation from biological reproduction to technologies already deeply bound up with reproducing human biology, even if the outcomes are not themselves biological matter. While xenofeminism foregrounds the "brute physicality" of techno-materialism (Hester 2018, 7), xenofeminist pursuits must reckon with machine learning as a technique indebted to the replication and reproduction of human physicality. This is especially critical since *Annihilation* not only evokes the way machine learning models, replicates, and reproduces human biology but also how it configures a "processed construct" (Cheney-Lippold 2018, 22) that uses biological features to (re)construct gender essentialisms.

For John Cheney-Lippold, an "algorithmic gender" (2018) emerges from the algorithmic processing of user data to construct a notion of gender suitable for online commercial exchange yet distinct from gender as a lived power construct. In this sense, gender is algorithmically produced as a construct that processes user information to fashion a "gender of profitable convenience" and designate particular attributes as "gendered" traits (Cheney-Lippold 2018, 12). In a similar vein, *Annihilation* demonstrates the way machine learning also entails a processed construct that situates biological traits as the grounds to essentialize gendered being as an outgrowth of biology. The film depicts scenes of two expedition teams that enter the Shimmer (even though we are told several expeditions have occurred). The first was a team of male military soldiers led by Kane (Oscar Isaac), Lena's husband, who early in the film returns home to Lena after he becomes the first of two people to return from the Shimmer (Lena ultimately being the other). The second is Lena's team of female scientists who encounter videographic documentation of the military team's experiences in the Shimmer. Where machine-learning technologies undergirding facial recognition software are "programmed to sort people into two groups—male or female" (Gault 2019), the Shimmer's inferences about the roles assumed by Lena and Kane analogize this sorting as one that situates biology as a determinant of gender.

When Kane's doppelgänger returns home, he resumes the social roles of husband and soldier even though the Shimmer only possesses the ability for biological reproduction, as the film provides no indication that it intuits social or cultural context. By mastering the biological traits of Kane, the Shimmer's reproduction of Kane's bodily form enables him to assume Kane's roles as husband and solider insofar as these roles are understood as genetically encoded. Likewise, when "Lena" emerges from the Shimmer, the doppelgänger has learned about her professional aptitude and domestic longing by "reading" its presence in DNA. For both "Kane" and "Lena," the reproduction of biological traits entails an implicitly essentialist reproduction of social roles—as can be read from their ability to seamlessly assume these roles. To be clear, this line of argumentation does not suggest that the roles of "scientist" or "soldier" are themselves retrograde or undesirable as pegged to gender; rather, the point is that the ability of machine learning to systematize and reify deterministic relationships between biology and designations of gender is antithetical to the antiessentialist aims of xenofeminism.

Viewed through this lens, machine learning not only categorizes among normative constructs of "male" and "female" in accordance with its programming, it also reproduces essentialist notions of gender as a constellation of innate inborn traits programmed into genetic hardwiring. Xenofeminism's wariness of biological reproduction collapsed into social reproduction (Hester 2018, 42–43) must therefore contend with the tendency of machine learning to reproduce human biology as automated technique and situate biological traits as the grounds to reconstruct and reinforce essentials notions of gender. With respect to the particulars of *Annihilation* as it relates to machine learning and xenofeminist antinaturalism, circling back to the overall aims of this chapter helps to place the contributions reaped from analyzing *Annihilation* in their broader contexts, as well as the contributions gleamed from analyses of *Covenant* and *Arrival*.

* * *

This chapter contributes to xenofeminist theory and application by arguing for tropes of recent extraterrestrial films as pathways to better conceptualize the risks of anchoring xenofeminist liberation to automated technologies and techniques. *Alien: Covenant* allegorizes technological autonomy as a condition prone to logics of domination and its relationship to capitalist expansion, as the xenomorph's autonomy operates at odds with the ability to pursue a postcapitalist world necessary to enact xenofeminist liberation. In *Arrival*, the nonlinear temporality of alien communication epitomizes the logic of calculation and anticipation undergirding algorithmic predictability, whereby the film's female linguist must ultimately conform to the inevitability of reproductive essentialism by presenting maternal pursuits as alienated from professionalized scientific pursuits across (a)temporal dimensions. In *Annihilation*, an alien ecosystem creates human doppelgangers by learning how to recognize and replicate human bodily attributes, just as machine learning ascribes social attributes and roles to women based on the recognition and replication of feminized bodily traits.

As a very recent intervention into political landscapes, opportunities abound for media scholars to build off the work in this chapter and more precisely address aspects of xenofeminist thought and application. While this chapter has analyzed extraterrestrial films to identify risks for xenofeminist alienation as a politics tied to automated technologies, additional critical inquiries can continue to deepen the conceptual and practical dimensions of xenofeminist thought, particularly given this book's focus on intersectional politics.

Thus, building off the work of this chapter and research applying theories of race and class to automated technologies (Noble 2018; Eubanks 2018), subsequent research could focus on the tendency of automated technologies to reinforce distinctions of class and race and, likewise, to more concertedly situate race, class, and other identify components alongside the antiessentialist and abolitionist aims of xenofeminism. Additionally, given socioeconomic motivations undergirding and engrained in technological design and practice, political economy approaches could illuminate the particularities of the "AI industry" (Dyer-Witheford, Kjosen, and Steinhoff 2019), the "social industry" (Sandvig 2015), "platform capitalism" (Srnicek 2017), and other digital economies primed with economic motivations and ideologies that intersect with xenofeminist concerns. Moreover, as emerging socialist paradigms continue to link egalitarian socioeconomic provisioning with automated technologies (Bastani 2019; Phillips and Rozworski 2019), the role of xenofeminist thought should be considered in light of how changing social relations to production might also address xenofeminist aims.

Finally, in the wake of the COVID-19 pandemic, opportunities abound to consider the relationships among xenofeminist politics, automated technologies, and the "regressive effects" (Madgavkar et al 2020) of burdens placed on women as a result of COVID-related politics. As very recent scholarship advocates for intersectional feminist approaches to inequities in COVID-related data governance (Berkhout and Richardson 2020; D'Ignazio and Klein 2020), socioeconomic notions of "value" (Ozkazanc-Pan and Pullen

2020), and unique challenges for women in the Global South (Al-Ali 2020), subsequent work could leverage insights gleaned from this essay to further interrogate the relationship between automation and alienation, especially given the dependency on a range of automated technologies to mitigate work responsibilities, interpersonal relationships, and health status from the domestic sphere. As many people find themselves indefinitely working from home amid the COVID-19 pandemic, the risk of a "new normal" (Bonacini et al 2021) for income inequality and gendered ideas around public and private labor warrant ongoing consideration as to whether or not alienation can be unmoored from capitalist exploitation and put to work (in the home and elsewhere) for gendered abolition. Additionally, the pervasiveness of algorithmic culture is one fruitful arena for interrogation, given recent indications about the gendered bias of algorithmic decision-making (Gutierrez 2021; Werner 2020) and the limitations of automated technologies as a liberatory tool from capitalist toil (Benanav 2020), especially with regard to speculations about an impending "wave" of automation as an outgrowth of task-efficiency measures implemented by businesses to keep capital flowing amid COVID-related shutdowns and economic disruption (Hinsliff 2020; Knight 2020).

NOTES

1. For a contrasting view of the potential of technology to advance post-gender plurality, see Bouma, chapter 10, this book. — Ed.

2. For a parallel critical take on understandings of technology as a path to transcending biological reproduction, this time in the context of transhumanism, see Lakshmanan, chapter 5, this book. — Ed.

REFERENCES

Al-Ali, Nadje. 2020. "Covid-19 and Feminism in the Global South: Challenges, Initiatives and Dilemmas." *European Journal of Women's Studies* 27, no. 4: 333–347.

Atanasoski, Neda, and Kalindi Vora. 2019. *Surrogate Humanity: Race, Robots, and the Politics of Technological Futures*. Durham: Duke University Press.

Bader, Christopher D., Joseph O. Baker, and F. Carson Mencken. 2017. *Paranormal America: Ghost encounters, UFO sightings, Bigfoot Hunts, and Other Curiosities in Religion and Culture*. New York: NYU Press.

Bastani, Aaron. 2019. *Fully Automated Luxury Communism*. London: Verso Books.

Baym, Nancy K. 2015. *Personal Connections in the Digital Age*. Hoboken: John Wiley & Sons.

Benanav, Aaron. 2020. *Automation and the Future of Work*. London: Verso.

Berg, Anne-Jorunn, and Merete Lie. 1995. "Feminism and Constructivism: Do Artifacts Have Gender?" *Science, Technology, & Human Values* 20, no. 3: 332–351.

Berkhout, Suze G., and Lisa Richardson. 2020. "Identity, Politics, and the Pandemic: Why is COVID-19 a Disaster for Feminism(s)?" *History and Philosophy of the Life Sciences* 42, no. 4: 1–6.

Bonacini, Luca, Giovanni Gallo, and Sergio Scicchitano. 2021. "Working from Home and Income Inequality: Risks of a 'New Normal' with COVID-19." *Journal of Population Economics* 34, no. 1: 303–360.

Brown, Bridget. 2007. *They Know Us Better Than We Know Ourselves: The History and Politics of Alien Abduction.* New York: NYU Press.

Carruthers, Anne. 2018. "Temporality, Reproduction and the Not-Yet in Denis Villeneuve's Arrival." *Film-Philosophy* 22, no. 3: 321–339.

Cheney-Lippold, John. 2018. *We Are Data: Algorithms and the Making of Our Digital Selves.* New York: NYU Press.

Creed, Barbara. 2015. *The Monstrous-Feminine: Film, Feminism, Psychoanalysis.* London: Routledge.

Dean, Jodi. 1998. *Aliens in America: Conspiracy Cultures from Outerspace to Cyberspace.* Ithaca: Cornell University Press.

D'Ignazio, Catherine, and Lauren F. Klein. 2020. "Seven Intersectional Feminist Principles for Equitable and Actionable COVID-19 Data." *Big Data & Society* 7, no. 2: https://doi.org/10.1177/2053951720942544.

Dyer-Witheford, Nick, Atle Mikkola Kjosen, and James Steinhoff. 2019. *Inhuman Power: Artificial Intelligence and the Future of Capitalism.* London: Pluto Books.

Eubanks, Virginia. 2018. *Automating Inequality: How High-Tech Tools Profile, Police, and Punish the Poor.* New York: St. Martin's Press.

Faulkner, Wendy. 2001. "The Technology Question in Feminism: A View from Feminine Technology Studies." *Women's Studies International Forum* 24, no. 1: 79–95.

Fisher, Mark. 2009. *Capitalist Realism: Is There No Alternative?* London: Zero Books.

Fleming, David H., and William Brown. 2018. "Through a (First) Contact Lens Darkly: Arrival, Unreal Time and Chthulucinema." *Film-Philosophy* 22, no. 3: 340–363.

Gault, Matthew. 2019. "Facial Recognition Software Regularly Misgenders Trans People." *Vice*, February 19, 2019. https://www.vice.com/en_us/article/7xnwed/facial-recognition-software-regularly-misgenders-trans-people.

Gillespie, Tarleton. 2014. "The Relevance of Algorithms." In *Media Technologies: Essays on Communication, Materiality, and Society,* edited by Pablo J. Boczkowski, Kirsten A. Foot, and Tarleton Gillespie, 167–194. Cambridge: MIT Press.

Gutierrez, Miren. 2021. "New Feminist Studies in Audiovisual Industries | Algorithmic Gender Bias and Audiovisual Data: A Research Agenda." *International Journal of Communication* 15, no 1: 439–461.

Halberstam, Judith. 1991. "Automating Gender: Postmodern Feminism in the Age of the Intelligent Machine." *Feminist Studies* 17, no. 3: 439–460.

Hester, Helen. 2018. *Xenofeminism*. London: John Wiley & Sons.

Hinsliff, Gaby. 2020. "The Next Wave of Coronavirus Disruption? Automation." *The Guardian,* May 1, 2020. https://www.theguardian.com/commentisfree/2020/apr/30 /coronavirus-disruption-automation.

Introna, Lucas, and David Wood. 2004. "Picturing Algorithmic Surveillance: The Politics of Facial Recognition Systems." *Surveillance & Society* 2, no. 2/3: 177–198.

Jones, Emily. 2018. "A Posthuman-xenofeminist Analysis of the Discourse on Autonomous Weapons Systems and Other Killing Machines." *Australian Feminist Law Journal* 44, no. 1: 93–118.

Jones, Emily. 2019. "Feminist Technologies and Post-Capitalism: Defining and Reflecting Upon Xenofeminism." *Feminist Review* 123, no. 1: 126–134.

Kavanaugh, James H. 1980. "Son of a Bitch: Feminism, Humanism, and Science in 'Alien.'" *October* 13: 91–100.

Kember, Sarah, and Joanna Zylinska. 2012. *Life After New Media: Mediation as a Vital Process*. Cambridge: MIT Press.

Knight, Will. 2020. "The Pandemic Is Propelling a New Wave of Automation." *Wired*, June 12, 2020. https://www.wired.com/story/pandemic-propelling-new -wave-automation/.

Laboria Cuboniks. 2015. "Xenofeminism: A Politics for Alienation." http://www .laboriacuboniks.net/20150612-xf_layout_web.pdf.

Madgavkar, Anu, Olivia White, Mekala Krishnan, Deepa Mahajan, and Xavier Azcue. 2020. "COVID-19 and Gender Equality: Countering the Regressive Effects." *McKinsey Global Institute.* www.mckinsey.com/featured-insights/future -of-work/covid-19-and-gender-equality-countering-the-regressive-effects.

Matzner, Tobias. 2019. "The Human is Dead—Long Live the Algorithm! Human-algorithmic Ensembles and Liberal Subjectivity." *Theory, Culture & Society* 36, no. 2: 123–144.

McKie, Robin. 2019. "Dr Alien, PhD: The Horror Classic that Academic Loved." *The Guardian*, March 24. https://www.theguardian.com/film/2019/mar/24/alien-horror -classic-that-academia-loves.

Melzer, Patricia. 2010. *Alien Constructions: Science Fiction and Feminist Thought*. Austin: University of Texas Press.

Microsoft Corporation. "Face API." https://azure.microsoft.com/en-us/services/co gnitive-services/face/.

Miyazaki, Shintaro. 2012. "Algorhythmics: Understanding Micro-Temporality in Computational Cultures." *Computational Culture*, 2. http://computationalculture .net/article/algorhythmics-understanding-micro-temporality-in-computational -cultures.

Mohri, Mehryar, Afshin Rostamizadeh, and Ameet Talwalkar. 2018. *Foundations of Machine Learning*. Cambridge: MIT Press.

Munster, Anna. 2011. *Materializing New Media: Embodiment in Information Aesthetics*. Hanover: Dartmouth College Press.

Newton, Judith. 1990. "Feminism and Anxiety in 'Alien.'" In *Alien Zone: Cultural Theory and Contemporary Science Fiction Cinema*, edited by Annette Kuhn, 82–87. London: Verso.

Noble, Safiya Umoja. *Algorithms of Oppression: How Search Engines Reinforce Racism.* New York: NYU Press.

Nye, David. "Technological Prediction: A Promethean Problem." In *Technological Visions: The Hopes and Fears that Shape New Technologies*, edited by Marita Sturken, Douglas Thomas, and Sandra Ball-Rokeach, 159–176. Philadelphia: Temple University Press.

Ozkazanc-Pan, Banu, and Alison Pullen. 2020. "Reimagining Value: A Feminist Commentary in the Midst of the COVID-19 Pandemic." *Gender, Work & Organization* (November): https://doi.org/10.1111/gwao.12591.

Paßmann, Johannes, and Asher Boersma. 2017. "Unknowing Algorithms: On Transparency of Unopenable Black Boxes" In *The Datafied Society: Studying Culture Through Data*, edited by Mirko Tobias Schafer and Karin Van Es, 139–146. Amsterdam: Amsterdam University Press.

Phillips, Leigh, and Michal Rozworski. 2019. *The People's Republic of Wal-Mart: How the World's Biggest Corporations Are Laying the Foundation for Socialism.* London: Verso.

Purse, Lisa. 2019. "Square-jawed Strength: Gender and Resilience in the Female Astronaut Film." *Science Fiction Film & Television* 12, no. 1: 53–72.

Sandvig, Christian. 2015. "The Social Industry." *Social Media+ Society* 1, no. 1: 2056305115582047.

Schudel, Matt. 2014. "H.R. Giger, Artist Who Designed the Creature in the 1979 Film 'Alien,' Dies at 74." *The Washington Post*, May 3, 2014. https://www.washingtonpost .com/entertainment/hr-giger-artist-who-designed-the-creature-in-the-1979-film -alien-dies-at-74/2014/05/13/6818becc-dabb-11e3-8009-71de85b9c527_story.html.

Srnicek, Nick. 2012. *Platform Capitalism.* London: John Wiley & Sons.

Sturken, Marita, Douglas Thomas, and Sandra Ball-Rokeach, eds. 2004. *Technological Visions: The Hopes and Fears that Shape New Technologies*. Philadelphia: Temple University Press.

Suchman, Lucy, and Jutta Weber. 2016. "Human-Machine Autonomies." In *Autonomous Weapons Systems: Law, Ethics, Policy*, edited by Nehal Bhuta, 75–102. Cambridge: Cambridge University Press.

Torry, Robert. 1994. "Awakening to the Other: Feminism and the Ego-ideal in Alien." *Women's Studies: An Interdisciplinary Journal* 23, no. 4: 343–363.

Trafford, James, and Pete Wolfendale, eds. 2020. *Alien Vectors: Accelerationism, Xenofeminism, Inhumanism.* Milton Park: Routledge.

Werner, Ann. 2020. "Organizing Music, Organizing Gender: Algorithmic Culture and Spotify Recommendations." *Popular Communication* 18, no. 1: 78–90.

Winner, Langdon. 1978. *Autonomous Technology: Technics-Out-of-Control as a Theme in Political Thought.* Cambridge: MIT Press.

Witten, Ian H., Eibe Frank, Mark A. Hall, and Christopher J. Pal. 2016. *Data Mining: Practical Machine Learning Tools and Techniques.* Morgan Kaufmann.

Wojciehowski, Hannah Chapelle. 2018. "When the Future Is Hard to Recall: Episodic Memory and Mnemonic Aids in Denis Villeneuve's Arrival." *Projections* 12, no. 1: 55–70.

Chapter 12

Simulation and Synesthesia in *Rez*

Virtual Reality and the Queer Erotechnics of Becoming-Machinic

tobias c. van Veen

FEELING THE MACHINE

The virtualization of the world continues apace with global events that tear us from the commingling of flesh. The unfolding climate catastrophes and pandemics of the twenty-first century have amplified the usage and development of virtual worlds and their interactive technologies beyond video gaming to sectors of education and health, governance, and labor. World-building technologies developed for video gaming have influenced other mediums, such as the use of the Unreal Engine to simulate interactive virtual worlds in film studios (Good 2020). This advanced level of virtualization has precursors in video gaming, notably *Rez* (United Game Artists 2001), which seeks to simulate, and stimulate, the synesthetic pleasures of the human body as it intersects with the machinic.

Writing on her blog *GameGirlAdvance* in 2002, video game studies scholar Jane Pinckard describes her first encounter with *Rez* on the PlayStation 2 as "a stunningly beautiful thing" (2002). *Rez* is a rail shooter video game in which electronic music melodies are assembled by the player as they target the game's abstract opponents. The sophisticated aesthetic design of *Rez* caught Pinckard's attention: "It is a 'music shooting' game, but of such elegance and coolness it's in a genre all its own" (2002). Video game producer Tetsuya Mizuguchi writes that *Rez* was unique enough that "not everyone understood what we were trying to do with the game" (2015). Yet the fans who "*really* got it" declared it "visionary. Revolutionary. Ahead of its time." What made *Rez* unique, says Mizuguchi, is its exploration of synesthesia: "We always listen to music by ear, and you can watch the visuals moving,

237

the dynamics in *Rez*, so it's kind of a cross-sensation feeling" (quoted in Bramwell 2006). *Rez* was prototyped as the "K Project," referencing Wassily Kandinsky's synesthetic and surrealist attempts to paint sound (U64 2008). The game's geometric abstraction can be situated in an ongoing remediation of modernism—in which "new media" reproduce, remix, and replace prior media (Bolter and Grusin 2000, 55)—that includes not only Kandinsky's remediation of sound into sight but Wagner's *Gesamkunstwerk*, insofar as *Rez* sought to simulate the "total (synesthetic) art" of Virtual Reality before the latter's commercial availability.[1]

Like any video game, *Rez* is a total assemblage of hardware and software. Crucial to this assemblage is the affective production of a queer gaming body through an *erotechnics* of becoming-machinic that operates in the realm of *feeling* the body as part of the machine.[2] By the machinic *totality* I mean the total gaming assemblage that encompasses the flesh, from the screen and its supporting software and hardware through haptic peripherals and vibratory stimuli that generate cross-sensory audiovisual cues. The flesh, I contend, becomes-machinic as a part of this circuit. One could also say that the intersectional aspects of the human body transect the process of becoming-automated. Turning to queer phenomenology (Ahmed 2006), I contend that becoming-machinic queers the body as it (re)circuits pleasure in the perpetual ludic play of the game's "queer narrative middle" (Chess 2016, 88). Queer narratives of play eschew traditional climax or catharsis just as they render porous distinctions between fleshy subject/technological object. Queer theory effectively reframes foundational debates in game studies over video gaming as ludology or narratology, suggesting instead that the erotechnic play of the former disrupts heteronormative expectations of the climax in the latter.

The specificity of the gaming body that is attenuated to *Rez* is cross-platform. Though I focus on the PlayStation 2 (2001), in which *Rez* was coupled with the Trance Vibrator, I also consider the discourse that envelops its general design and reception, focusing upon the ways in which *Rez* was claimed by its designers to *stimulate* synesthesia in the flesh—as well as *simulate* synesthesia in its visual and audio elements. I situate *Rez* as a sociotechnical assemblage that articulates the affective potential of stimulating/simulating synesthesia within gaming environments to the production of speculative concepts that address this same milieu. One such concept is synesthesia itself, which besides its standard meaning—cross-sensory perception such as "seeing sounds" or "hearing touch"—signifies the ambiguity of how the phenomenological event of synesthesia is registered, and represented, in a language that would be undone by its effects. Synesthesia is not a given, and though it might reference an empirical or neurological event, it enters into the discourse of *Rez* as an expectation, while at the same time naming an affective intensity that parallels the discourse of pleasure.

In the synesthetic environment of *Rez*, music and sound are prioritized in a way that video game studies and phenomenology alike often neglect. The gamer advances through the ludic narrative of *Rez* by targeting visual elements that interplay with audio cues, shaping a rhythmic soundtrack of electronic music. Its sound affects the body in not just audible registers but in its tactile force as vibration. As Steve Goodman suggests (2010), it is because of its pre-sensory vibration that Deleuze and Guattari considered sound as "piloting" the phenomena of synesthesia (2000, 348). In *Rez*, haptic feedback is provided through the controller, but it is its optional and scarce haptic accessory, the Trance Vibrator, which as a black, oblong vibrating peripheral, presents an (un)ambiguously erotic dimension to the game's attempt to simulate, and stimulate, synesthesia. Yet what constitutes synesthesia, and how it is simulated and yet stimulated in *Rez*, calls for an analysis of the circuit—affective and technological—between the gamer's body and the assemblage of the game. The vibration of this circuit, I contend, is *erotechnic*. At the intersection of *eros* and *tekhnē*, erotechnics harnesses the affective chaos of synesthesia, particularly in its vibratory force as tactility and sound, to stimulate a becoming-machinic that transgresses simulation. But the simulation itself is crucial, and *Rez* simulates two synesthetic environments, one irreal (cyberspace) the other surreal (rave culture). The distinction between irreal and surreal signifies the difference between a space of simulation by way of hypothetical representation (the cyberspace irreality of the *simulation of* synesthesia) and the simulation of a space of *stimulation* by way of collective participation (the surreality of rave culture's synesthetic environment). These distinctions will be unfolded below.

Consensual Hallucinations of Cyberspace: From *Tron* and *Neuromancer* to *Rez*

Seemingly weightless as it dances in space, the graceful body of the *Rez* avatar is illuminated by a neon grid. The avatar's resemblance to the Users of the 1982 film *Tron* (Lisberger 1982) is remarked upon by Pinckard, who connects the game's imaginary to its science fictional antecedents:

> The premise places you as a futuristic hacker, avoiding security systems and navigating through databases to crack codes in a way strongly reminiscent of *Tron*. Your mission: information needs to be free. (Pinckard 2002)

Like *Tron*'s Users, who have shed their flesh to inhabit the lightworld of the computer mainframe in a battle against the Master Control Program (Lisberger 1982), the aim of *Rez* is to manipulate the audiovisual Grid. As radiant geometric shapes assemble across the screen, the player must target as many as possible to construct a coherent audiovisual narrative that advances the game. The visual language of *Rez* operates at the intersection of its cyberspace imaginary to its flow of targeting. Indicative of how the game draws the body into its flow, in *Rez* "the gamer *releases* the trigger to shoot" (Wark 2007, 127). The hands are kept tense until fingers are released to target the game's visual icons. Like the limited virtuosity required to play digital music videogames, such as *Guitar Hero* (2005; see van Veen and Attias 2011), the movement of the fingers becomes rhythmic through automated quantization, their timing constitutive to the soundtrack's assembly. The rhythmic gestures of trigger *release* constitute the first level of erotechnics in *Rez*'s audiovisual gameplay. The rhythmic caressing of the controller initiates the pleasurable release of sensory overload that *Rez* aims for.

With the completion of Grid assemblies, a string of computer code descends from the top-left corner, evoking the stream of connectivity text from a dial-up internet connection. Though it may be that "storyline is the bad faith of the game" (Wark 2007, 142), one of the aesthetic appeals of *Rez* is how it simulates the science fictional imaginary of cyberspace. Playing *Rez* is as much becoming a User in *Tron* as it is "jacking in" to the Matrix, as imagined in Gibson's *Neuromancer*, where "Shadows twisted as the holograms swung through their dance. [. . .] Lines of light ranged in the nonspace of the mind, clusters and constellations of data" (1984, 37; 51). *Rez* visualizes what many thought cyberspace would turn out to be. By simulating future technological imaginaries that have not yet—or might never —come to pass, *Rez* undertakes a nostalgic simulation of a lost future while, at the same time, it tries to model its emergence.

Feeling the Vibe: Synesthesia, Rave Culture, and *Rez*

> At first, I was thinking of a game for people who liked club music, something they could enjoy without actually going to the club. (Jun Kobayashi, *Rez* Game Director 2001)

If the science fictional imaginary of cyberspace is one component of *Rez*'s simulated content, the actuality of rave culture's synesthetic and hallucinatory events is the other. Writing on her blog, Jane Pinckard describes *Rez* as part *Tron*, part rave:

> "Flying down neon corridors shooting space/machine beings in time to techno music," as Justin [Hall] put it, is a pretty good description, as is "*Tron* on Ecstasy." The game bills itself as synesthesia—union of senses. But even that doesn't begin to relay the feeling of the game. Even without the trance vibrator, the game puts you into a trance state—it's a raver's game, a game of pure sensation. (Pinckard 2002)

Rez simulates one part science fictional irreality, one part rave culture sur-reality. The game's developers commissioned the recombinant soundtrack from well-known electronic music producers including Japanese techno producer Ken Ishii (the thematic "Creation the State of Art (Part 1–6)"), U.K. breaks producer Adam Freeland, U.K. hip-hop remix duo Coldcut (with Tim Bran), and German sound-artist Oval, a.k.a. Markus Popp. The beta version of *Rez* featured unauthorized music from techno outfit Underworld (apparently inspiring the title), the abstract beats of Richard D. James a.k.a. Aphex Twin, and the electronica of The Chemical Brothers (see U64 2008). Rave culture's electronic musical styles were seen as integral to the development of *Rez*, insofar as its rhythms helped shape the motion and flow of the gameplay by providing a coordinating soundtrack. The element of audiovisual coordination here is crucial: it is the 4/4 structure of electronic music that coordinates the micro-gestures of play to the game's production of visual narrative. Furthermore, it is the synchronization of a player's micro-gestures to the music, much like dance, that actively constructs the simulation of rave's synesthesia. Without music, *Rez* becomes all but unplayable; silencing its soundtrack has the effect of not only removing vital cues as to when a player should be triggering the controller but of eliminating the audio feedback that occurs when targeting. Contrary to standard game design, in which music and sound effects are secondary programming concerns to a game's visual environment, in *Rez* music is positioned as central to game development and play. In *Rez*, the coordinated interaction of visual and haptic feedback with music is crucial for stimulating synesthesia in game players by simulating both surreal (rave) and irreal (cyberspace) environments.

Rave is a radically synesthetic environment. Its basic elements of repetitive electronic music, played over a high-volume soundsystem with various modes of visual stimuli, are aimed at inducing a trance state. The sensory chaos of a rave engenders the trance state of "temporary self-effacement" (Gaillot 1999, 55), with one's sense-of-self becoming lost or merged to the milieu. Wark observes a similar phenomenon in *Rez*, wherein its "pulsating, phosphorescing quality [. . .] gives the gamer a feeling of a particularly intense loss of self" (2007, 137). In both cases, the loss of self is experienced through the intensity of embodiment as it intersects a technical apparatus that stimulates synesthesia. Reports of synesthesia, aided by the "body technologies" of empathogens (Rietveld 1998), populate the literature of Electronic Dance Music Culture (EDMC) studies. The empathogenic drug MDMA, used to treat Post-Traumatic Stress Disorder, is a popular recreational drug because it enhances the senses while producing feelings of well-being and empathy (Saunders and Doblin 1996). Simon Reynolds writes of how "MDMA turned out to have a uniquely synergistic/synesthetic interaction with [electronic] music," writing likewise of "electronic listening music" that "synesthesia was a common aesthetic goal" (1999, 83; 199). These "synesthetic effects" are amplified so that "sounds seem to caress the listener's skin," to the point where "sound becomes a fluid medium in which you're immersed" (Reynolds 1999, 84).

Deleuze and Guattari argue that sound has a "piloting role" in the phenomena of synesthesia: colors are "*superposed* upon the colors we see, lending them a properly sonorous rhythm and movement," much like the audible phenomena of sound itself (2000, 347–48). Sound owes this power to its "machinic phylum that [. . .] makes it a cutting edge of deterritorialization" (2000, 348). Bodies are deterritorialized—abducted, disorganized, energized, and ungrounded—by the machinic phylum of sound: "sound invades us, impels us, drags us, transpierces us. It takes leave of the earth" (2000, 348). Precisely because of its "strong deterritorialization," however, sound rebounds with territorial ferocity, amplifying tendencies toward the "potential fascism of music" and its "numbing" effects (Deleuze 2000, 348), such as in military marches, consumer jingles, rave's tendency toward mindless hedonism, and videogaming's tendency toward behavioral addiction. Rave and *Rez* alike are mechanisms of repetition, expressed in their routinizations of gesture that correlate to a soundtracking of electronic rhythm. While *Rez* simulates the audiovisual elements of repetition that define rave culture, it crucially does not catalyze the dancing body of the raver. Instead, its flesh-body remains supine, its movements constrained to the micro-gestures of the gamer. This difference is crucial to differentiating their respective mobile intensities, though it may not be for differentiating their respective abilities to stimulate synesthesia.

In *Sonic Warfare*, Steve Goodman draws upon a number of observations concerning sound and synesthesia in EDMCs as well as policing and military applications, suggesting that sound is primarily amodal, as "the abstract, cross-mediality of rhythm" that ontologically precedes the distribution of the senses (2010, 47). The affective chaos of synesthesia is not a clinical pathology but "foundational of the affective sensorium" as the very condition of possibility for the specificity of the senses (Goodman 2010, 47). Keeping in mind that *Rez* attempts to channel sound to image in a "cross-sensory" play, it too deploys the sonic strategies of repetitive electronic music to amplify or intensify the machinic phylum of synesthesia that lurks behind all single-sensory perception.

In rave culture, the qualitative effect of machinic synesthesia is an ephemeral but entirely material condition of collective synchronicity known as the *vibe*. In the cybernetic feedback loop of the rave's soundsystem to its dancefloor, affective chaos engenders collective movement, wherein "bodies become one with the music, each distinct gesture the fractal-expression of the singular sonic algorithm" (Hemment 1996, 31). The qualitative or rather immersive and affective state of a rave can be observed through what Graham St John calls its "radical empiricism of the vibe" (2013). The *vibe* is that vibratory assemblage of bodies to rhythmic sound in a display of gestural *communitas* (Turner 1969), with the aim of achieving what Hemment calls "ek-stasis," a "reciprocal relation wherein the musical flow is actualized as body-music: music becoming embodied in dance, dancers becoming disembodied as music" (1996).

The successful gameplay of *Rez* is remarkably similar. With its neon avatar immersed in a luminescent world of flashing lights and beset by hallucinatory abstractions that pulse in time to the recombinant techno compositions of Ken Ishii, playing *Rez* simulates the experience of raving—though without moving anything other than fingers to participate in the sonic algorithm—in which selecting visual symbols corresponds to the gameplay's interactive composition of electronic music melodies.[3]

The key to both raving and gaming *Rez* is decoupling the segregation of the senses and feeling the vibe. In both scenarios, one attempts to become what the "Trance Vibrator" signifies in its naming—that becoming in which the body rhythmically vibrates in an ephemeral and pleasurable state of synesthetic trance.

REZ AND THE BECOMING-MACHINIC OF ABSTRACT GENDER

If *Rez* simulates both irreal and surreal environments of synesthesia, it does so through modeling the synesthetic body by way of its avatar.[4] In *Rez*, the avatar appears human only in its modeling of an ideal geometric form. For many of the game's Areas, the avatar is feminine in form, with curved hips and breasts, suggesting the ongoing representation (and exploitation) of the female subject and yet, in light of its Platonic geometricism, the avatar offers a sensate challenge to phallogocentric ideals of pure mathematics. The avatar is never static, always in becoming, always moving, providing a succession of wireframe models for player identification. There is an aesthetic beauty to the avatar, insofar as it models what a Virtual Reality experience of "becoming" that avatar might feel like. The discourse of "feeling" is significant. Producer and creator Tetsuya Mizuguchi (2015) emphasizes it throughout his writings on *Rez*, along with his stated intention to explore "synesthesia" through Virtual Reality (see Bramwell 2006).

Consumer VR was not yet ready for *Rez*'s original release in 2001. Fifteen years later, *Rez Infinite* (Q Entertainment, Monstars Inc., Enhance Games 2016) launched on the PlayStation 4, with the option of Sony's PlayStation VR headset. The *Rez* avatar will cease to animate the modeling of the VR body, and instead players will become one. The full impact of VR *Rez* is yet to come. Even so, clinical research using Virtual Reality to treat depression by psychologists Caroline J. Falconer, et al. (2014) suggests that player identification with a synchronized (and aesthetically pleasing) VR body poses "mental health benefits." Slater has found that "although they know it's not true, people nonetheless have a strong perceptual illusion that this [VR] body is their own body" (quoted in Young 2016). It is here that VR research and *Rez* intersect. The initial editions of *Rez* attempted to accentuate identification with the avatar, from the participatory constructivism of the game's audio-visual architecture to the haptic feedback of the erotically charged Trance Vibrator. Just as the *Rez* vibrator seems to have been explicitly engineered for female bodies, it's worth noting that the clinical VR research was conducted on "healthy female volunteers high in self-criticism to experience self-compassion from an embodied first-person perspective within immersive virtual reality" (Falconer et al. 2014). The dual use of the gendered female body in both gaming and clinical VR speaks to the need for a critical yet speculative convergence of approaches from feminist technoscience, video game studies, queer/affect theory, and process philosophy.[5] It is also here that intersectional approaches to the human body require a new vector: the automation of the body by way of virtualization. We need to ask what *kind* of body, and *whose* body, and *how* this body is programmed, constructed, and perceived in VR. Such critique extends to its "beneficial" mental health effects if disconnection from the pleasing VR body returns the gamer to that geometrically imperfect sticky body of flesh left behind, beset by a troubled world depleted of the neon aesthetic pleasures of *Rez*.

In *Rez*, the body is not always gendered. Nor is it necessarily human. Nor is it even alien *qua* species. In Area 01, the avatar is a hexagonal spheroid, whereas in Area 05, no body presents itself but a geometric abstraction. *Rez* director Jun Kobayishi explains that "when you devolve, when you're almost dead," the avatar turns into a spheroid while "for every 8 blue items you collect, your character evolves" (2001). From Areas 01 through 03, what reveals itself as an ungendered though "feminine" avatar must be shaped, piece by piece, by selecting its body components. The abstraction of the avatar to Platonic geometric form is coupled with the dismembered avatar at various stages of *Rez*, suggesting a further Enlightenment trope whereby the reassembly and representation of "gendered" physical attributes arise from the level of geometric abstraction. Such ambiguous gender abstractions do not, however, defer the "rescue the princess trope," as *Rez* projects a female AI as the object of rescue. Yet such rescue appears in *Rez* as anticlimactic, a ruse that revels in the "dilatory moments of the denial of climax," as Chess contends of *Super Mario Bros* (2016, 89). The mutability of the avatar is the first key to understanding how *Rez* calls to attention the "feeling" of the gamer's body. "Feeling" the body as becoming-avatar also renders visually affective how the "[queer] pleasure of video game narrative is about becoming, rather than about coming" (Chess 2016, 88). As I explore below, *Rez* gameplay recombinates the avatar to music in a perpetual circuit of ludic pleasure that likewise queers the player's body.

Queering the Senses

In addressing feeling, it is important to underscore how synesthesia addresses gender as queering of subject/object. Haraway's celebrated cyborg ontology commences from such a thesis (2004). The synesthetic trance of ludic play—the ephemeral vibe of a dancefloor or a *Rez* gamer session—is fundamentally queer. First, it delays climax to that "never-ending [queer] middle that is dependent on alternative pleasures" (Chess 2016, 88; see Roof 1996). Second are the alternative pleasures themselves: the ludic play of the queer middle transgresses subject/object through cross-sensory becoming-machinic. As synesthesia decouples and recouples the senses—precisely as their amodal condition—it likewise undoes the heteronormative axes of subject/object, bringing into affective chaos the supposed neutrality of the "upright, straight, and in line" body (Ahmed 2006, 159). As the senses unstraighten, a cross-sensory "confusion or transgression" takes place that cross-circuits the gendering of subject/object as active (male)/passive (female). Taken as the praxis of inciting affective chaos, synesthesia deconstructs the orientation ontology of being-in-line. The dancefloor has long been a queer space, from disco to rave culture (Lawrence 2003). Yet, like videogaming, it also induces queering as becoming-machinic. Feeling gender as an-other becomes fluid in the cross-sensory flow of affect that cross-circuits technological object to fleshy subject.

Affect theory is helpful here to thinking "the queer potential of all gaming" (Chess 2016, 92) as it gestures beyond binarism to the *virtual* (Massumi 2002), that field of potential multiplicity signaled in Deleuze and Guattari's term, "*becoming*-machinic." Distinct from emotion or feeling, affect names "pre-individual bodily forces augmenting or diminishing a body's capacity to act" (Clough 2010, 207). Massumi writes that the feelings or emotions of the subject, as the narrative of consciousness, is "subtracted" from the "virtual remainder," or excess of affect (2002, 25–9). Playing with the affective chaos of synesthesia, I suggest, renders sensate the *queer* multiplicity of the *virtual*—that "intense mode of abstraction into which sensation potentially infolds" (Massumi 2002, 98). Here, "the virtual" is not akin to Virtual Reality but signals sensate potentiality of non-binary multiplicity. It is VR, as technology, that attempts to infold the virtual through cross-sensory queering of the player. This is why the synesthesia of raves, playing *Rez*, or strapping into

VR can feel disorientating: dis-orientation disrupts the perceived neutrality of the straight subject and induces a queer becoming-machinic. In its infolding of cross-sensory queerness, the body becomes disorientated from the biological orientation of sex but also from being-subject. In Sara Ahmed's queer phenomenology, addressing Fanon (2008), queering occurs at "the point at which the body becomes an object" (2006, 159). Embracing the body-object as radical praxis of disorientation takes place in various decolonial and antiracist contexts, including Caribbean philosophies of creolization (Chude-Sokei 2016) and the "radical black ontogenesis" of becoming-alien/ machinic that underscores Afrofuturism (van Veen 2013). Admittedly such praxes raise concerns over undoing dehumanization by way of becoming the object. In Haraway's "ironic political myth" of the cyborg, the latter is said to supersede race, gender, and class even as it somewhat ironically remains an allegory of "category disputations over the black body in America" (Chaney 2003, 265). As I argue elsewhere (van Veen 2013), the subtext to declaring "I, robot" remains racialized enslavement, not the least by virtue of its invention as a metaphor for slavery in Karel Čapek's 1920 play *R.U.R.* That white supremacy, enslavement, and racialization persist is cause to reiterate concerns over who controls and benefits from the potential manipulation of disorientation. Science fiction abounds with near-future dystopias where video gaming is the digital means of enslavement. The immersive and synesthetic "feelings" of *Rez* amplify Clough's concern with "those technologies that are making it possible to grasp and to manipulate the imperceptible dynamism of affect" (2010, 207). Yet it is not necessarily the case that to oppose manipulation one needs to "return to the subject as the subject of emotion," as Clough argues (2010, 53). The self can be conceived as a radically resistant object, precisely in its resistance to an imposed subjectivity—a point made by Fanon in his phenomenology of racial objectification (2008). As much as becoming-machinic risks the disastrous irony of playing unwittingly at an ontological allegory of racialized enslavement, when mobilized as queer praxis it troubles the conditions of the latter, as Ahmed's reading of Fanon suggests. Ahmed's radical praxis of queer disorientation signals the *potential*—good or bad—of circuiting fleshbodies to technological bodies, a position reflected in the cybernetic theses of Stiegler (1998), Derrida (2000), Lacan (2002) and Kittler (1997).

As queer synesthetic play, disorienting the "feeling" of the body becomes equitable on the machinic phylum with its other(s)—often technological/ machinic, science fictional, and alien bodies—in a radical decentering of human privilege. It is the queer posthuman potential of the latter that *Rez* entertains. *Rez* attempts to simulate these domains and avatars of exhuman becoming. When stimulated in gameplay, becoming-machinic disorientates the felt experience of the (human) body itself. Such "moments of disorientation," writes Sara Ahmed, "are vital" (2006, 157): they are life-giving, precisely because the disorientated body refuses to be shaken to its senses as but a (human) being-in-line.

Addiction and Attraction/Melding and Touch

The intersection of virtual and fleshy bodies is the site—and the event—of the becoming-machinic body in *Rez*. Becoming-machinic is a queering, as fleshbodies disorientate to cyborgian desires. What needs to be considered is how the fleshbody is "attracted" and "addicted," by way of "feeling"—particularly in the vibratory registers of sound and touch—to the *erotechnics* of becoming-machinic.

Martti Lahti (2003, 159) has studied the "bodily, almost noncognitive, dimension" of being "sucked into the computer" and how it contributes to what Ted Friedman describes as a "meditative state, in which you aren't just interacting with the computer, but melding with it" (1999, 137). Such a melding, even if of the cognitive variety—science fiction posits the exact finger placements of a Vulcan Mind Meld—requires touch. Fingers caress and tap controllers. Controllers vibrate against sweaty flesh. It is at the intersection of cognitive melding and affective touch that we encounter what Lahti and Friedman both describe as the "attractive and addictive potential" of videogaming (Lahti 2003, 159). It is here that queer becoming-machinic is reterritorialized through the numbing effects of addictive repetition. Yet, while the pleasure received from becoming-machinic explicates the attraction, and thus risks the addiction of videogaming, this same pleasure arouses its circuit of queering—its *erotechnics*.

Rez stimulates attraction through its attempt to both simulate and stimulate synesthesia, a desiring assemblage attractive precisely because it transgresses the boundary between fleshbody and machine. Lahti, commenting upon Anne Balsamo (1996), notes that "Games—in particular fighting, shooting, and racing games—are a symptomatic site of a confusion or a transgression of boundaries between the body and technology" (2003, 158). But where is "the body" located? *Who* or *what* is (the) body? In the affective circuit of becoming-machinic, it is *vibration* that complicates body policing of subject/object. Vibration, as the haptic force of affect, brings into close relation both things and sounds that touch "the body." But as the flesh is touched so is the thing. Soundsystems and vibrators alike stimulate (erotic) responses as touchy-feely components. Cybernetics already suggests that the difference between fleshbody and machine is but a difference between two information feedback loops (Aarseth 1999; Wiener 1988). Between subject and object would be a difference of one *technics* to another. The pleasure circuit of video gaming signals a queering (of) erotechnics.

The attractive potential of video gaming is its erotechnics of the other. It is the queer feeling of attraction to the machinic other—the very process of (erotic) vibration with the assemblage of avatar, controller, vibrator, darkened room, stimulation of eyes/ears/flesh, and fellow gamer(s)—that constitutes the erotechnics of video gaming. Erotechnics signals how pleasure need not take place between two Subjects. The Trance Vibrator suggests an erotechnical circuit that operates through the sensate vibration of rhythm in a queering assemblage composed of human and nonhuman actors. Such leveling of human exceptionalism, insofar as erotechnics inscribes pleasure among nonhuman actors, echoes a point made by Kittler, who argues that the death of the Subject is more or less a technico-epistemic effect of the invention and dissemination of Alan Turing's Universal Discrete Machine (1997, 135; 145).

SIMULATION AND STIMULATION OF SYNESTHESIA

Rez attempts to simulate synesthesia at the same time that it attempts to stimulate it. Between the two is an intriguingly blurry boundary whereby simulated effects potentiate stimulated affects. Before turning to *Rez*'s tactile methods of stimulation, I wish to question this metaphysical boundary. The queering effects of erotechnics, I suggest, play in the undecidability of who/what stimulates—the simulation or the real of synesthesia.

As Eskelinen and Tronstad point out of gaming assemblages, "for every individual system we also have, to some degree, an individual medium" (2003, 198). Besides its surreal (rave) and irreal (cyberspace) imaginaries, the medium that *Rez* is trying to model, or indeed simulate, is that of Virtual Reality, which in the 1990s was "bulky, blurry, and slow" (Mizuguchi 2015). In the twenty-first century, VR is a consumer phenomenon consisting of visual goggles and headphones, coupled with glove controllers.[6] In VR mode for 2016's *Rez Infinite*, writes Mizuguchi, "the game comes closest to what we saw in our heads when we were creating [the original version]: vivid colors that blend seamlessly into one another, crystal-clear textures, and razor-sharp lines at full 1080p HD, all swimming around you at the speed-of-life 60 frames per second (120 frames per second in PS VR), with full 3D audio (PS VR) or 7.1 surround sound (PS4) that, well, truly surrounds you" (2015).

The specific effect of a VR system—which is to say the "feeling" of its medium, its sensory immersion wherein senses mingle—is its attempt to induce synesthesia. The language of VR speaks to the feeling of being surrounded by sensate technology. Wearing a VR headset and handling feedback-enhanced controllers, the machinic assemblage attempts to eliminate the distraction of any sensate exteriority. VR technologies attempt to increase the quantifiable resolution of simulated affect—in frames per second, digital audio sample rates, color spectra resolutions, and visual pixel counts—to the point where its matrix becomes indistinguishable from the sensate world of irreducible affective quality. The philosophical distinction between quantities and qualities becomes blurred at a mere 120fps.[7]

Since its first edition, *Rez* has pursued affective immersion through the simulation of synesthesia. "Simulation" is a useful term in this context, insofar as it is an "alternative to representation and narrative" (Frasca 2003, 223). It is an alternative not because simulation does not represent but because simulation undertakes a different modality of representation insofar as it seeks to supersede it. Simulation strives for the replica of (a) representation. In Platonic metaphysics, it constitutes a second-order representation as it signifies upon the signifier and not the referent. Baudrillard (2006) argues that simulation can be severed from the order of representation entirely, gaining its autonomy *qua* simulacra.

Though synesthesia is the stated intention of *Rez*, as a "cross-sensation feeling," it arises from the perception of (a) simulation. As Frasca points out, simulation models the "behaviors" of an object, and in this case, that object is VR itself. What all versions of *Rez* prior to 2016 produce is a simulation of VR's hypothetical environment of synesthesia. This environment is no longer hypothetical, though it is imperfect, and signals a simulacra to-come. What the Trance Vibrator suggests, however, is that the erotechnics of the gaming assemblage is possibly crucial to stimulating, and not just simulating, synesthesia. The perfection of simulacra requires stimulation through erotechnics, as gaming supersedes play to become life itself:

> The only thing important about that technology for *Rez Infinite* is how it makes you *feel*; hopefully, it makes it easier to forget the fact that you're sitting in front of a TV playing a game, and instead lets the real world melt away into a swirl of incredible sights and sounds that could only ever exist in your imagination. That's exactly how we want players to feel—that they aren't just playing *Rez Infinite*, they're living it. *Rez Infinite* is the futuristic "synesthesia" experience I've wanted to create from the beginning. (Mizuguchi 2015)

What Mizuguchi imagines here is properly an improper *event*. At the blurring of gameplay to life irrupts an evental intersection: Kurzweil's singularity of artificial intelligence and modernism's attempts at *Gesamkunstwerk*. It is worth briefly returning to how *Rez* draws inspiration from Kandinsky's synesthetic attempts to paint sound. Wolf sees the video game as an "extension of abstract art" (2003, 47), insofar as it focuses on the temporality of an *event* rather than its representation—a point that resonates with *Rez*'s erotechnical gameplay of "the queer middle" (Chess 2016, 89). Yet what constitutes an event, when its modus operandi is simulation? As Baudrillard points out, it is the specificity of the event itself that is at stake in simulacra—even though we might say its becoming constitutes one (2006).

The signifiers I deploy to describe *Rez* and its effects—from "fleshbody" to "erotechnics"—all gesture at the affective intensity of an event that struggles from the synesthetic ludic play of videogaming technology into language. The discourse around *Rez* engenders such an expectation of the synesthetic event. This expectation awaits an affective intensity that shatters language, leaving it struggling to discern what simulates and what stimulates. In the ambiguity between the two is a queering of the senses that trembles the boundaries of a language that would separate the *who* from the *what*. It is at this intersection that *Rez* positions the Trance Vibrator.

The Ecstasy of Abstract Flesh

Rez heightens the vibe of simulated VR. It does so by "cross-sensation": the cross-wiring of visual cues to sonic rhythms in an attempt to gamify the simulation of synesthesia. Yet *Rez* adds a further tactile stimulant, the limited-edition Trance Vibrator peripheral. With the Trance Vibrator, *Rez* shifts from simulating synesthesia—in its simulation of the User's body as VR avatar and its gameplay sensorium of sight to sound—to stimulating its disoriented affects through a third dimension: the haptic. With the Trance Vibrator, *Rez* introduces the "primal scene" of erotechnics, where the discourse of the circuit irrupts into the tactile and queer pleasures of becoming-machinic.

As Pinckard and her partner Justin Hall began playing *Rez*, they plugged in the Trance Vibrator, an oblong black box that "started thumping like crazy in time with the music" (2002). Exploring the "sublime visual and aural experience," heightened by the game's "invincible 'travelling' mode," which allows the player to "just sit back and move through the levels without worrying about your avatar's taking damage," Pinckard tuned into the erotic dimension of the oblong object:

> Well, what would you have done? I moved the vibrator into my lap.
>
> "Oh my GOD! This game rocks! Here, you play." I handed him the controller but you'd better believe I kept that vibrator right there in my lap.
>
> Justin hadn't been paying that much attention to the exact location of the vibrator, since the game is visually quite absorbing, too. But I wanted to concentrate a little more on the, er, physical aspect of the game. I took off my pants.
>
> Justin got the idea. "Wow."
>
> We sat side by side on our makeshift couch, I with the trance vibrator and Justin with the controller. As the levels got more advanced, so did the vibrations . . . revving up to an intense pulsing throbbing . . .
>
> "Oh, God!" (Pinckard 2002)

At this point, the vibe becomes as tactile as it is machinic. Synesthetic rhythms coalesce in the technological object caressing flesh. Deities are invoked in pleasurable ek-stasis. The exterior world is shuttered: "We drew the curtains and darkened the room. Justin played for hours," Pinckard recounts. The game becomes secondary to the synesthetic immersion in a tactile yet virtual reality.

> Pretty soon the levels and the images onscreen were just a faint blur to me. I knocked off my glasses and leaned back. I was in a daze. From far away, it seemed, I could hear Justin saying things like, "I made it to the next level!" and "This is cool!" but I was lost in my own little trance vibrating world. (Pinckard 2002)

In her blog post, Pinckard discusses the erotic dimensions of the Trance Vibrator, questioning whether it was the designer's intent. According to interviewer Tom Bramwell, Mizuguchi read Pinckard's account and had this to say: "'That was my idea,' he says, grinning, when asked about it. 'That was kind of a joke, but a very serious joke. No sexual meaning,' he continues" (in Bramwell 2006). Jokes are, of course, always a serious affair, as any introductory reading of Freud tells us (1981). Jokes displace anxieties while signifying unconscious desire—which is to say, they spill some secrets about the discourse of the circuit. But the irony of Mizuguchi's joke is that he is entirely correct: indeed, no sexual *meaning* was intended. Nor could it be. For Mizuguchi appears to have taken his stated intention of machinic synesthesia entirely seriously. At play is the erotechnical, not the sexual.

When asked by Bramwell where the vibrator is "meant to go when you're playing," Mizuguchi responded, "I like to feel the vibration by the foot. So I think it's the feet and the hands. It's a good balance. Some people bite them! I think that's really dangerous, actually" (in Bramwell 2006). Given that if one is playing the hands are occupied, the possible locations for sensate vibrations are somewhat limited. The oral use of a vibrator can be as stimulating as feet or genitalia. All are fleshy sites of erotic contact. The feet are the source of some 7,000 nerve endings, as well as the site of Freud's analysis of the fetish,[8] and for Mizuguchi, an evident site of pleasure for a podophile.

Yet with two players, wherever the Trance Vibrator is placed, the touch of one player is circuited to the flesh of the other via the peripheral. But the Trance Vibrator is no mere conduit. It participates as a technological other. When placed on the fleshbody, it disturbs the orientation of sex by inducing the erotechnic pleasure of becoming-machinic. As Hall remarks, "It was a bit odd. My fingers were working the controls, but they were also kind of working you." This strangeness accounts for the fact of the stranger: Pinckard and Hall are engaged in a queer machinic three-way. Hall asks, "I wonder how it works on guys?" (2002), suggesting that the cross-sensory queerness of their machinic partner transgresses its straight overcoding as a phallic vibrator for female bodies. The machinic queering of *Rez* decouples hetero-normativity while disrupting the "conservative beliefs about heterosexuality and 'proper' romance" that, according to Mia Consalvo, many games reify (2003, 171)—not the least because the third partner is the queer hardware of the game itself.

At the close of her blog post, Pinckard reveals that this is not the first time she's "used a game component to, er, stimulate [her]self physically" (2002). The creative misuses of vibratory game controllers suggest a vastly understudied queer erotechnics of gamer culture. Discussing the merits of various controllers—none of which had been appropriately shaped nor pro-grammed—Pinckard describes how the Trance Vibrator is perhaps the first machinic partner intended for erotechnical pleasure in a synesthetic system:

> That's why I was so excited by *Rez's* trance vibrator, since it seems to have no other purpose than to act as a masturbatory aid. Its shape is pretty nice, it can slip easily under your skirt or in your panties, it comes with a protective "glove" which you can wash, and it emits a regular pulsating rhythm that gets ever more intense and thrilling the deeper you go into the game. Damn, by the end I was writhing on the floor! Synesthesia indeed. (Pinckard 2002)

Mizuguchi says that "engaging all your senses was also the idea behind the original Trance Vibrator" (2015). The Trance Vibrator materializes the game as an-other, an object not unlike a hand, a foot, or a mouth, able to interact with any fleshy concentration (or orifice) with nerve endings.

Presented by Mizuguchi at the 2015 PlayStation Experience launch of *Rez Infinite*, "the ultimate version of the Trance Vibrator" is the "Synesthesia Suit." Outfitted in glowing neon, the Synesthesia Suit appears to envelop the body with the sensations of *Rez* while transforming the player into a visual representation of the game's avatar. Consisting of twenty-six actuators, the suit delivers complex haptic sensations to a player's arms, legs, and torso. Unlike the Trance Vibrator, the suit will only be available to "one of the lucky few" as a "promotional item" for the PlayStation Experience installation. But this too suggests another inside joke. Mizuguchi writes that you "really need to *feel* [it] for yourself," going on to say that "the experience of playing *Rez Infinite* in the suit truly brings players inside the game and its music" (quoted in Hurwitch 2017). Though consumer-grade VR is here for the masses, fully immersive Virtual Reality—which in a capitalist dystopia can be imagined as the ultimate escape, severing the User entirely from the chaotic world through a pleasurable and orgasmic immersion in the cyberpunk nostalgia for the Net that never was—is only available for the queer autoaffection of the one (to the other). The ultimate Freudian joke would be that one such suited body will be the First Contact with AI. *Rez* implicates nothing less: to communicate with the machinic other, at the apogee of Kurzweil's singularity, will require the queer erotechnics of touch. Not climactic, but anticlimactic, in the circuiting of pleasure. For how else will the machinic other know if the fleshbodies in its surrounds are anything more than a simulation of its programming?

CONCLUSION: ERGODIC ART, BECOMING-MACHINIC, AND QUEER EROTECHNICS

Ergodic phenomena are produced by some kind of cybernetic system, i.e., a machine (or a human) that operates as an information feedback loop, which will generate a different semiotic sequence each time it is engaged. (Aarseth 1999, 32–33)

As we play around with *Rez*, it becomes clear that new conceptual lexicons are required to discuss the emergence of VR as it passes from the domain of simulated synesthesia to its actualization. These lexicons likewise need to be attentive to the reciprocity and affective intensity that takes place in cybernetic circuits of (post)human assemblages that, in turn, destabilize clear distinctions between the *who* and the *what*. Earlier elaborations of video gaming as an ergodic art, under the rubric of ludology, remain useful here as they attend to cybernetic models that complement Chess's call to think about video gaming as queer narratology (2016). Though *Rez* is a case example for ludology where, as Frasca argues, games are "not held together by a narrative structure" (2003, 222), such a perspective depends upon a heteronormative model of narrative climax. *Rez* retains the "perverse" narrative of the queer middle that delays climax to the plateau of pleasure. Furthermore, it enacts its queer narrative middle through an erotechnics of gameplay that harnesses the synesthetic potential of ergodic art into a queer becoming-machinic of the gamer. The ergodic also offers a cybernetic framework in which to analyze the discourse of the circuit between the gamer and the gamed. *Rez* is one of few video games that attempts to actualize what the cybernetic model entails: that the human body is but a component, a becoming-machinic body within an emergent assemblage of gameplay. When combined with the simulation of synesthesia—a simulation that confuses or transgresses its effects through stimulation when coupled with haptic devices that render queer the senses—the cybernetic model is in need of new concepts to explicate the convergence of *eros* and *technics*.

The straight logic of narrative gameplay—to win or to lose, advance or restart, for the avatar to live or die (see Roof 1996)—is short-circuited by the queer undecidability of a body's affective potential. This includes technological bodies. The queer synesthetic feedback circuit of machine<>flesh can be summarized in the concept of *erotechnics*. Erotechnics names the cybernetic feedback circuit in which bodies become-machinic through the queering of straight senses into synesthesia and its pleasures.

Erotechnics has particular effects, then, on how games are categorized and differentiated from other aesthetic regimes. Eskelinen and Tronstad argue that video games are distinct from traditional art, wherein the static work requires interpretation, while ergodic art requires nontrivial labor from the user, such as participatory narratives (2003, 198). Ergodic videogaming, however, consists of "pleasurable systems and modes that are not dominated by interpretative interest," wherein the user must interpret the gameplay so as to "proceed from the beginning to the winning or some other situation" (2003, 199). Calling to mind Chess's "alternative pleasures" of the queer narrative middle, *Rez* crucially shifts the concept of "winning" to "another situation" that defers the heteronormative teleology of climax. With erotechnics, the very outcome of a game shifts to the in-folding of feeling becoming-machinic n+1. Becoming-machinic takes place at the level of what Clough reads as the "circuit from affect to emotion" (2010, 207), where the sensation of affective chaos particular to becoming-machinic catalyzes a queering of the senses through the dis-orientation of the subject<>object.

Becoming-machinic constitutes a kind of event. Events are not climaxes, but in this instance constitute a disorientation of straight subject–object relations. For Eskelinen and Tronstad (2003), videogames are akin to the Happenings of Allan Kaprow, performance art in general, and thus for Wolf, constitute a participatory technological activity that is "time-based, interactive, and event-based" (2003). The event of becoming-machinic portends a qualitative potential to the erotechnic circuit—the queer infolding of the virtual—that has yet to be fully explored, particularly for its communicative role with artificial intelligence.

At the same time, and unlike transhumanist ideologies that seek virtual transcendence through shedding of the body, *Rez* undertakes a synesthetic exploration of what Lahti calls a "new cyborgian relationship with entertainment technologies" (2003, 158).[9] This entertaining relationship is erotechnical. The cyborg, as flesh sutured with technological prosthetics, knows itself as cyborg because it touches its other-self with technology. Erotechnics articulates *technics* to *eros* through the information loop of flesh to machine bodies in which information is vibration. It attends to how the erotic operates through a technological feedback circuit in which the body exhibits addiction and attraction, in a technico-libidinal economy of ergodic play. And its evental effects are by no means constrained to gaming, though it is erotechnical gaming that models the event of machinic intelligence to-come.

ACKNOWLEDGMENTS

Many thanks to Nathan Rambukkana for publishing work that tugs at the boundaries of technics and thought, to Hillegonda Rietveld for feedback on previous iterations, and to Jane Pinckard for stimulating the conversation.

NOTES

1. Further, such remediation of modernist artistic synthesis, abstraction, and collage is double, insofar as modernism itself remediated prior artistic forms into its avant-garde manifests of "total art."

2. For a parallel discussion of becoming-machinic, but with respect to how processes of gamification can entrain the body to algorithmic control under neoliberalism, see Gómez, chapter 3, this book. — Ed.

3. In a further twist of remediation (Bolter and Grusin 2000), today's electronic music production and DJ software appears to remediate, in part, the videogaming conventions of *Rez*, as both the sonic arts of djing and composition become "gamified" through the use of screen-based software and brightly-lit haptic controllers (see van Veen and Attias 2011, 2012).

4. Images that complement this article have been archived at: http://quadrantcrossing.org/rez.

5. To name but a few texts: Sarah Ahmed's work on queer phenomenology (2006); Anne Balsamo's interrogation of supplementary technological bodies (1996); Donna Haraway's feminist/queer cyborg manifesto (2004); N. Katherine Hayles' critique of cybernetics and the lost bodies of posthumanism (1999); the launch of *Ada: A Journal of Gender, New Media, and Technology* by Carol Stabile and Kim Sawchuk (2012); and reflections on gender and internet identity by Sherry Turkle (1995).

6. Initial products include the HTC Vive and the Oculus Rift (Shanklin 2016).

7. Film theorists would say 24fps, including Kittler. But they haven't seen a thing yet.

8. Freud's phallogocentric theory claims that foot fetishes are the result of castration anxiety (1991).

9. For a critical discussion of transhumanism that touches on the desire to transcend the limits of human bodies, see Lakshmanan, chapter 5, this book. — Ed.

REFERENCES

Aarseth, Espen. 1999. "Aporia and Epiphany in *Doom* and *the Speaking Clock*: Temporality in Ergodic Art." In *Cyberspace Textuality*, edited by Marie-Laure Ryan, 31–41. Bloomington and Indianapolis: University of Indiana Press.

Ahmed, Sara. 2006. *Queer Phenomenology: Orientations, Objects, Others*. Durham: Duke University Press.

Balsamo, Anne. 1996. *Technologies of the Gendered Body: Reading Cyborg Women.* Durham: Duke University Press.

Baudrillard, Jean. 2006. *Simulacra and Simulation.* Translated by Sheila Fraser Glaser. Ann Arbor: University of Michigan Press.

Bolter, Jay David, and Richard Grusin. 2000. *Remediation: Understanding New Media.* Cambridge: MIT Press.

Bramwell, Tom. 2006. "'Non-Sexual' Rez Trance Vibrator Was My Idea— Mizuguchi." *Eurogamer.net*, July 27, 2006. http://www.eurogamer.net/articles/n ews250706mizuguchi.

Chaney, Michael. 2003. "Slave Cyborgs and the Black Infovirus: Ishmael Reed's Cybernetic Aesthetics." *MFS: Modern Science Fiction Studies* 49, no. 2: 261–83.

Chess, Shira. "The Queer Case of Video Games: Orgasms, Heteronormativity, and Video Game Narrative." *Critical Studies in Media Communication* 33, no. 1: 84–94. doi:10.1080/15295036.2015.1129066.

Chude-Sokei, Louis. 2016. *The Sound of Culture: Diaspora and Black Technopoetics.* Middletown: Wesleyan University Press.

Clough, Patricia T. 2010. "The Affective Turn: Political Economy, Biomedia, and Bodies." In *The Affect Theory Reader*, edited by Melissa Gregg and Gregory J. Seigworth, 206–28. Durham: Duke University Press.

Consalvo, Mia. 2003. "Hot Dates and Fairy-Tale Romances: Studying Sexuality in Video Games." In *The Video Game Theory Reader*, edited by Mark J.P. Wolf and Bernard Perron, 171–94. New York: Routledge.

Deleuze, Gilles and Félix Guattari. 2000. *A Thousand Plateaus: Capitalism and Schizophrenia.* Translated by Brian Massumi. Minneapolis: University of Minnesota Press.

Derrida, Jacques. 1997. *Of Grammatology.* Translated Gayatri Chakravorty Spivak. Baltimore: Johns Hopkins University Press.

———. 2000. "Signature Event Context." In *Limited Inc*, translated by Alan Bass, 1–24. Evanston: Northwestern University Press.

Eskelinen, Markku, and Ragnhild Tronstad. 2003. "Video Games and Configurative Performances." In *The Video Game Theory Reader*, edited by Mark J.P. Wolf and Bernard Perron, 195–200. New York: Routledge.

Falconer, Caroline J., Mel Slater, Aitor Rovira, John A. King, Paul Gilbert, Angus Antley, and Chris R. Brewin. 2014. "Embodying Compassion: A Virtual Reality Paradigm for Overcoming Excessive Self-Criticism." *PLoS ONE* 9, no. 11. doi:10.1371/journal.pone.0111933.

Fanon, Frantz. 2008. *Black Skin, White Masks.* Translated by Richard Philcox. New York: Grove Press.

Frasca, Gonzalo. 2003. "Simulation Versus Narrative." In *The Video Game Theory Reader*, edited by Mark J. P. Wolf and Bernard Perron, 221–35. New York: Routledge.

Freud, Sigmund. 1955. "A Note Upon the Mystic Writing Pad." In *Standard Edition of the Complete Psychological Works of Sigmund Freud*, translated by James Strachey, Vol. Xix, 227–32. London: Hogarth Press.

———. 1991. *On Sexuality.* Translated by James Strachey. London: Penguin.

————. 1981. *Jokes and Their Relation to the Unconscious.* Translated by James Strachey. New York: W. W. Norton.

Friedman, Ted. 1999. "Civilization and Its Discontents: Simulation, Subjectivity, and Space." In *On a Silver Platter: CD-ROMs and the Promises of a New Technology*, edited by Greg M. Smith, 132–50. New York and London: New York University Press.

Gaillot, Michel. 1999. *Multiple Meaning: Techno, an Artistic and Political Laboratory of the Present.* Translated by Warren Niesluchowski. Paris: Éditions dis Voir.

Gibson, William. 1984. *Neuromancer.* New York: Ace Books.

Good, Owen S. 2020. "How Lucasfilm used Unreal Engine to Make The Mandalorian." *Polygon*, 20, 2020. https://www.polygon.com/tv/2020/2/20/21146152/the -mandalorian-making-of-video-unreal-engine-projection-screen.

Goodman, Steve. 2010. *Sonic Warfare.* Cambridge: MIT Press.

Haraway, Donna. 2004. "A Manifesto for Cyborgs: Science, Technology, and Socialist Feminism in the 1980s". In *The Haraway Reader*, 7–46. New York: Routledge.

Hayles, Katherine N. 1999. *How We Became Posthuman: Virtual Bodies in Cybernetics, Literature, and Informatics.* Chicago: University of Chicago Press.

Hemment, Drew. 1996. "'E is for Ekstasis'." *New Formations* 31: 23–38.

Hurwitch, Nick. 2017. "Go to Synesthesia: Writing on *Rez.*" *Medium*, January 23, 2017. https://medium.com/@heWIZARD/go-to-synaesthesia-writing-on -rez-ee0b4349b1f9.

Jeter, K. W. 1998. *Noir.* New York: Spectra.

Kittler, Friedrich. 1997. *Literature, Media, Information Systems.* Amsterdam: Overseas Publishers Association.

Kobayashi, Jun. 2001. "Rez—2001 Developer Interview with Director Jun Kobayashi." *Shmuplations.com.* Accessed January 9, 2016. http://shmuplations.com/rez/.

Kurzweil, Ray. 2006. *The Singularity is Near: When Humans Transcend Biology.* New York: Penguin.

Lacan, Jacques. 2002. *Écrits: A Selection.* Translated by Bruce Fink. New York: Norton.

Lahti, Martti. 2003. "As We Become Machines: Corporealized Pleasures in Video Games." In *The Video Game Theory Reader*, edited by Mark J. P. Wolf and Bernard Perron, 157–70. New York: Routledge.

Lawrence, Tim. 2003. *Love Saves the Day: A History of American Dance Music Culture, 1970–1979.* Durham: Duke University Press.

Lisberger, Steven, dir. 1982. *Tron.* USA: Walt Disney Productions.

Lovink, Geert. 2002. *Dark Fiber: Tracking Critical Internet Culture.* Cambridge: MIT Press.

Massumi, Brian. 2002. *Parables for the Virtual: Movement, Affect, Sensation.* Durham and London: Duke University Press.

Mizuguchi, Tetsuya. 2015. "Rez Infinite Revealed for PlayStation VR." *PlayStation. Blog*, December 5, 2015. http://blog.us.playstation.com/2015/12/05/rez-infinite -revealed-for-playstation-vr/.

Pinckard, Jane. 2002. "Sex in Games: Rez+vibrator." *GameGirlAdvance.com*, October 26, 2002. http://www.gamegirladvance.com/2002/10/sex-in-games -rezvibrator.html.

Reynolds, Simon. 1999. *Generation Ecstasy: Into the World of Techno and Rave Culture*. New York: Routledge.

Rietveld, Hillegonda. 1998. *This is Our House: House Music, Cultural Spaces and Technologies*. Avebury: Ashgate.

Roof, Judith. 1996. *Come as You Are: Sexuality and Narrative*. New York: Columbia University Press.

Saunders, Nick and Rick Doblin. 1996. *Ecstasy: Dance, Trance & Transformation*. Oakland: Quick American Archives.

Scott, Ridley, dir. 1982. *Bladerunner*. USA: Warner Bros.

Shanklin, Will. 2016. "Why We Think the Oculus Rift is Still the Best VR Headset (Hands-on)." *Gizmag.com*, January 9, 2016. http://www.gizmag.com/best-virtual -reality-headset-oculus-rift-review-hands-on-2016/41264/.

Spinoza, Baruch. 1994. *A Spinoza Reader: The Ethics and Other Works*. Translated by Curley, Edwin. Princeton: Princeton University Press.

St John, Graham. 2008. "Trance Tribes and Dance Vibes: Victor Turner and Electronic Dance Music Culture." In *Victor Turner and Contemporary Cultural Performance*, edited by Graham St John, 149–73. New York: Berghahn.

———. 2013. "Writing the Vibe: Arts of Representation in Electronic Dance Music Culture." *Dancecult: Journal of Electronic Dance Music Culture* 5, no. 1. doi:10.12801/1947-5403.2013.05.01.11.

Stabile, Carol, and Kim Sawchuk. 2012. "Introduction: Conversations Across the Fields." *Ada: A Journal of Gender, New Media, and Technology* 1, no. 1. doi:10.7264/N3RN35SV.

Stiegler, Bernard. 1998. *Technics and Time, 1: The Fault of Epimetheus*. Translated by Richard Beardsworth and George Collins. Stanford: Stanford University Press.

Stone, Alluquère Rosanne. 1992. "Will the Real Body Stand Up? Boundary Stories About Virtual Cultures." In *Cyberspace: First Steps*, edited by Michael Benedikt, 81–118. Cambridge: MIT Press.

Turkle, Sherry. 1995. *Life on the Screen: Identity in the Age of the Internet*. New York: Simon & Schuster.

Turner, Victor. 1969. *The Ritual Process: Structure and Anti-Structure*. Chicago: Aldine.

U64 Staff & Contributors. 2008. "K Project (REZ) [Dreamcast—Beta / Prototype / Unused." *Unseen64: Beta, Cancelled & Unseen Videogames*. https://www .unseen64.net/2008/04/10/k-project-rez-prototype/.

van Veen, tobias c. 2013. "Vessels of Transfer: Allegories of Afrofuturism in Jeff Mills and Janelle Monáe." *Dancecult: Journal of Electronic Dance Music Culture* 5, no. 2: 7–41. doi:10.12801/1947-5403.2013.05.02.02.

van Veen, tobias c., and Bernardo Alexander Attias. 2011. "Off the Record: Turntablism and Controllerism in the 21st Century (Part 1)." *Dancecult: Journal of Electronic Dance Music Culture* 3 (1): http://dj.dancecult.net/index.php/journal /article/view/104/131.

van Veen, tobias c., and Bernardo Alexander Attias. 2012. "Off the Record: Turntablism and Controllerism in the 21st Century (Part 2)." *Dancecult: Journal of Electronic Dance Music Culture* 4 (1). http://dj.dancecult.net/index.php/journal /article/view/121/144.

Wark, McKenzie. 2007. *Gamer Theory.* Cambridge: Harvard University Press.

Wiener, Norbert. 1988. *The Human Use of Human Beings: Cybernetics and Society.* New York: Da Capo Press.

Wolf, Mark J. P. 2003. "Abstraction in the Video Game." In *The Video Game Theory Reader*, edited by Mark J. P. Wolf and Bernard Perron, 47–65. New York: Routledge.

Young, Nora. 2016. "A Virtual Treatment for Real Depression." *CBC Radio Spark with Nora Young*, March 13, 2016. http://www.cbc.ca/radio/spark/313-virtual-reality -actual-loneliness-and-more-1.3481377/a-virtual-treatment-for-real-depression -1.3482062.

LUDOGRAPHY

Hexadrive. 2008. *Rez HD.* [Xbox 360, Xbox Live Arcade], Seattle, United States: Microsoft Game Studios and Q Entertainment, played December, 2015.

Q Entertainment. 2011. *Child of Eden.* [Xbox 360 and PlayStation 3], Montréal, Canada: Ubisoft, played December, 2015.

———. Monstars Inc., Enhance Games. 2016. *Rez Infinite.* [Sony PlayStation 4, PlayStation VR, PlayStation Network, Microsoft Windows], Tokyo, Japan: Enhance Games, played April, 2016.

United Game Artists. 2001. *Rez.* [Sega Dreamcast and PlayStation 2], Tokyo, Japan: Sega, played December, 2015.

Index

ableism: misogyny and, 3, 106–7;
transhumanism and, 112. *See also*
disability
abolition: of (human) rights, 131; of
gender, 213, 220, 224, 228, 232–33
Actor Network Theory (ANT), 131
affect: affected public, 43, 142; affective
(chaos (sensory), 238–39, 245, 254;
machine (robots), 172–73, 184, 238;
manipulation, 62–63, 67, 70); haptic
touch, 247, 250; programmed, 152;
simulated, 248–50; sound, music,
239, 243; theory, 244–46. *See also*
Massumi, Brian
Ahmed, Sara, 71, 238, 245–47. *See also*
phenomenology (queer)
algorithm(s): advertising discrimination,
17–33; definitions and processes,
225–27; to detect plagiarism, 81–85;
Facebook, 21–23; overview of
scholarship, 1–3; sonic, 242–43. *See
also* algorithmic; machine-learning
algorithmic: control, 60, 68; culture, 5,
39, 42, 233; detection, 81; gender,
215, 230; identities, 3; justice, 67;
media, 41–43, 53, 55–56; platform,
59, 69; society, 18; support, 79
alien(s), 8–9; in science fiction (*Alien*
films), 213–18, 220–21, 231–32

alienation, 9; xenofeminist, 213–14,
224, 227, 232–33
artificial intelligence (AI), 79, 249;
in education and higher-learning,
77, 81; gender and, 195, 254;
posthumanism and, 121; social
robots and, 173; transhumanism
and, 101
artificial wombs, 102, 107–10, 114–15,
115n5. *See also* in-vitro fertilization
(IVF); robots (sex)
Ascender Alliance, 112–14. *See also*
Terasem; transhumanism
Asimov, Isaac, 159, 160
automation: alienation and, 233; of
the body, 62, 244; discussions of,
77–79, 195; of education, 81, 86,
88; gamification and, 60, 62–65;
post-industrial, 214, 217; robots
and, 135
autonomy: human, 217; posthuman
bodies and, 205–6; queer bodies
and, 200, 207; reproductive,
224; simulation and, 249; social
robots and, 173–74, 182–83;
technofeminist critique of, 219. *See
also* Haraway, Donna; women's
rights and, 127, 155; xenofeminism
and, 223, 231–32

261

Bates, Stephen, 197–98
Baudrillard, Jean, 249. *See also* simulation
becoming: machinic, 238–39, 243, 245–47, 250–51, 253–54; posthuman, 120, 200, 206–8, 214, 243; queer (through seduction, 198, 206–8; through the machinic, 238, 245–47, 253–54). *See also* Deleuze and Guattari
behavior: algorithms and, 52; behavioural (addiction, 242; psychology and control, 60–71); human norms of, 171
Bina48 (AI robot), 113–14, 124–25. *See also* Rothblatt; Terasem
bioethics: Christian and Jewish theological roots of, 101; eugenics and, 105. *See also* ethics
biopolitics. *See* biopower; democratic transhumanism and, 114; four camps of transhumanist, 101–5, 111; soft, 3, 52–53
biopower: soft, 52–53
Braidotti, Rosi, 69–71, 171, 199–201, 205–8. *See also* posthumanism
Breazeal, Cynthia, 171–73, 179. *See also* Kismet
Butler, Judith, 197–98, 204. *See also* gender

Campaign Against Sex Robots, 3, 150–53. *See also* Richardson, Kathleen
Campaign to Stop Killer Robots, 3, 156, 162n1. *See also* Campaign Against Sex Robots
capitalism: climate and, 128; libertarian, 108; life beyond (post-), 215, 217, 220–24; platform, 232; techno-, 60; xenofeminism and, 219–20. *See also* xenofeminism
capitalist realism, 222. *See also* Fisher, Mark
carebots, 8; as caretakers (for aging populations), 174–77; ethics of, 179–84; as healthcare robots, 169–70; long-term relationships with humans, 172–74, 178–79, 184–85; as potential social actors, 171–72. *See also* robots (social)
Cheney-Lippold, John, 3, 5, 39–40, 51–53, 56n1, 230. *See also* algorithms
COVID-19; conspiracy theories and, 67, 70, 72n2; digital disparities and, 24, 179–80; digital education and, 92n1; gender discrimination and, 232–33; post/transhumanism and, 115n5, 121; sex robots and, 161, 163n9; social robots (carebots) and, 169, 175–76, 182–84
Crenshaw, Kimberlé, 112, 121, 124, 170, 180. *See also* intersectionality
critical race theory, 124. *See also* race
Cuboniks, Laboria, 213–14, 219–20. *See also* xenofeminism
cyberfeminism, 199. *See also* xenofeminism
cyborg: disability and, 4; disembodiment and, 199; imaginaries, 202–5; manifesto, 219, 255n5; ontology, 9, 245–47, 254; queer, 8, 208. *See also* cyberfeminism; Haraway, Donna; robot

Data (*Star Trek* android), 149
Deleuze, Gilles (and Félix Guattari), 64, 239, 242, 245
Department of Housing and Urban Development (HUD), 17
Deterding, Sebastian, 59, 61, 65. *See also* gamification
digital disparities, 170, 179–80
disability: Americans with Disability Act, 179; digital (ad) discrimination and, 17, 20; essentialist model of, 106, 112; feminist, 107, 110; rights and justice movements, 5, 22, 112, 163n9, 179; studies, 4, 64, 115n2, 124. *See also* ableism

disembodiment: of information, 199; SOPHIE (electronic musician), 113, 202; transhuman fantasy of, 108, 113
disidentification: digital embodiment and, 204; (queer) posthumanist, 207; as specific mode of (queer) performance, 201, 204–5. *See also* Muñoz, José Esteban

embodiment: artificial stereotypes of, 2–3, 7; fluid, 113; material and human, 71; posthuman, 122–24, 201, 204; queer, non-binary and trans, 8–9, 195–99, 205, 207–8; robot (Sophia), 141; SOPHIE (electronic musician), 202; synesthetic, 241. *See also* disembodiment
erotechnics, 8–9, 209n3, 238–40, 247–50, 252–54. *See also* phenomenology (queer)
Ess, Charles, 156, 183. *See also* ethics
ethics: posthuman, 200; roboethics, 132, 180, 185; sex robotics and, 157. *See also* Richardson, Kathleen; of the singularity, 4; of transhumanist technologies, 101–15. *See also* bioethics
Eubanks, Virginia, 3, 18, 232. *See also* algorithms
Ex Machina (film), 4, 151, 161–62

Facebook, 5, 63; algorithmic advertising discrimination and, 17–24; critiques of, 24–31, 138; faking age and, 39–40, 42, 44–56. *See also* social media
feminism, 103; faux and pseudo, 155; intersectional and materialist, 60, 70–71; posthumanist, 111, 113–14, 124; techno-. *See* xenofeminism. *See also* intersectionality
Ferrando, Francesca, 99–100, 103, 122, 125, 129, 132. *See also* posthumanism
Fisher, Mark, 222. *See also* capitalist realism

Fourth Industrial Revolution, 1, 6, 153
Fuchs, Christian, 17, 21, 28, 64
Fukuyama, Francis, 123–24
fun: alternative fantasies of embodiment and, 204; behavioral control, gaming and, 6, 64–66; Gamergate and, 66–68; gamewashing and, 68–72; gaming, surveillance and, 59–60; SOPHIE and, 208
Futurama (animated science fiction show), 151, 157, 162

Gamergate, 66–67
game studies, 64, 66, 237–38, 244. *See also* video game(s)
gamewashing, 6, 60, 65–71
gamification: behavioral control and, 61–64; definition of, 60–61; design, 69; of society, 68. *See also* automation
gender: abolition. *See* abolition (gender); abstract, 243–44; algorithmic, 215, 230, 233; avatars and, 39–40, 44, 244–45; discrimination and (algorithmic) targeting, 17–18, 21–24, 28, 47, 50–56, 60; disparities and inequalities, 106, 141–45, 170; emancipation, 216–20; essentialism, 9, 101–2, 111–13, 214–15, 225, 228, 230–31; performativity and embodiment, 8, 113, 195–98, 200, 202, 206–7, 246; robots and, 154, 157, 160, 164n12. *See also* robots (sex); Sophia (robot), 7, 119, 145; stereotypes, 155. *See also* intersectionality; sexuality
Gesamkunstwerk (total art), 238, 249
Greene, Shelleen Maisha, 113–14. *See also* race
Guzman, Andrea, 1–2, 171–74, 184. *See also* Human-Machine Communication (HMC)

Halberstam, Jack (Judith), 124, 199–200, 218
Hall, Melinda, 101–11. *See also* feminism

Hamari, Juho, 61, 66, 69. *See also*
    gamification
Hansen, Mark, 199. *See also*
    phenomenology
Haraway, Donna, 171, 198–99, 219,
    245–46. *See also* posthumanism
Hayles, N. Katherine, 122, 125, 199,
    216, 228, 230, 255n5. *See also*
    posthumanism
health disparities, 179–80, 185. *See also*
    digital disparities
Her (film), 4, 151, 159, 161
Hester, Helen, 213–14, 217, 224, 230–
    31. *See also* xenofeminism
Hughes, James, 100–101, 111–14. *See
    also* posthumanism
humanism: discussion and critique of,
    7, 99–102, 108, 110, 112, 114, 125,
    132; rights and Enlightenment, 126,
    130. *See also* posthumanism
Human-Machine Communication
    (HMC), 1–2, 78, 172–74. *See also*
    Trance Vibrator
Humans (television show), 4, 151, 158

imaginaries, 196, 200; cyborg, 202;
    future technological, 240; queer
    posthumanist, 207–8; surreal and
    irreal, 248
incel (movement), 155–56, 163
Instagram, 40, 143, 205. *See also* social
    media
intersectionality: definition and
    discussion of, 60, 70, 112, 121,
    124, 170, 180; intersectional
    (algorithmic biases, 60;
    antidiscrimination statutes, 30;
    approach to public health, 184–85,
    232; datafication of identities, 42;
    framework in game studies, 64, 67,
    71, 238, 244; posthuman rights,
    121–24; transhumanist women,
    112–13, 115). *See also* Crenshaw,
    Kimberlé

intimacy: digital intimacy studies, 1; of
    posthuman becoming, 208; queer,
    204; sex robotics and, 152, 157
in-vitro fertilization (IVF): COVID-19
    and, 115n5; definition and critiques
    of, 102; social justice and, 112, 114–
    15; transhumanist ethics of, 105–6,
    109–10, 115n2. *See also* artificial
    wombs; misogyny
Ishii, Ken (electronic musician), 241,
    243
Istvan, Zoltan, 102, 107–12. *See also*
    transhumanism

Japan: as leading in carebots, 169, 174–
    75; techno music from, 241

Kandinsky, Wassily, 238, 249. *See also*
    synesthesia
Kaprow, Allan, 254. *See also*
    performance (art)
Kismet, 172–73. *See also* robots (social)

Latour, Bruno, 41. *See also* Actor
    Network Theory (ANT)
Levy, David, 7, 149–53, 156–63n5. *See
    also* robots (sex)
long-term relationships: with social
    robots, 171–84

MacCormack, Patricia, 197–201, 207.
    See also queer (theory and studies)
machine-learning, 139, 215; algorithms,
    5, 31, 41–42, 228, 231; care robots,
    173–74; definitions and discussion
    of, 227–31; Facebook, 21
Massumi, Brian, 245. *See also* affect
    (theory)
McGonigal, Jane, 67. *See also*
    gamification
MDMA (Methylenedioxy-
    methamphetamine), 241
mediation: queering of technological,
    196–97, 206. *See also* remediation

misogyny: anti-Black racism and, 112–14; Gamergate and, 66; Humanism and, 100; transhumanism and, 102, 104, 106–7, 109–14

modernism: objectification of women, 108; videogaming and, 238, 249, 255n1

Muñoz, José Esteban, 201–5. *See also* queer (theory)

Noble, Safiya Umoja, 2, 18, 32, 39, 43, 59, 232. *See also* algorithm(s)

performance: art (feminist), 8, 196, 200–206, 254; machine learning and, 228; performing bodies, 200, 230; social robotos and, 174; xenofeminism and, 218, 220

phenomenology, 197, 199, 255n5; queer, 238–39, 246

Pinckard, Jane, 237–40, 250–52. *See also* game studies

posthuman(ism): critical (ethics), 125, 130, 200–201; definition and discussion of, 121–24, 199–200; queer, 196, 202, 207–8, 246; rights, 119–20, 126–32; and transhumanism, 99–100, 102, 110, 114. *See also* transhumanism

preimplantation genetic diagnosis (PGD): COVID-19 and, 115n5; definition and critique of, 102; eugenic framework of, 106; social justice and, 114–15; transhumanist ethics of, 105, 109–10, 112. *See also* artificial wombs; in-vitrio fertilization (IVF); misogyny

privilege: human, 1–3, 5, 69, 100, 121, 246; systemic, 68–69, 101; women's, 151–52

queer: embodiments and identities, 8–9, 70, 113, 197–98, 201, 205–8, 219; gaming, 71, 238, 245–48,

250–53; middle, 245, 249, 253; phenomenology, 238, 246. *See also* Ahmed, Sara; posthumanism, 196, 200–202, 207–8, 246; queering (as becoming-machinic, 245–48, 252; (posthuman) embodiment, 196–97, 201, 207; gaming, 71; the senses, 245, 250, 254); (sex) robots, 110, 113, 250–53; theory and studies, 64, 71, 159, 195, 197, 238, 244

race: algorithms and, 40, 52, 70; digital discrimination and, 17–23, 28–29, 60, 180; (post/trans)human, 104, 113; robots and, 7, 154, 170, 246; xenofeminism and, 213, 232. *See also* intersectionality

Rancière, Jacques, 127–28

rave (culture), 9, 239–45, 248

remediation: of modernism and sound, 238, 255n1, 255n3

Rez (video game), 237–39, 244–45, 249. *See also* video game(s)

Richardson, Kathleen, 2–3, 7, 109, 150–57, 162–63. *See also* Campaign Against Sex Robots; ethics

rights: civil, 20, 26, 28–32, 157; disability, 5, 22, 163n9; Indigenous and ecological, 128–29; (post)human, 113, 119–23, 126–32, 151 (critiques of, 123–26); sex worker's, 151, 153; women's, 126–27, 136–37, 140–41, 171. *See also* posthuman(ism)

robosexuality: definition and discussion of, 7–8, 150–51, 154, 162n3; sci-fi visions of, 157–62. *See also* robots (sex)

robot(s), 78, 135; carebots, 124, 169–72, 173–78, 184–85; discussion of kinds, 3–8; ethics and, 179–84; posthuman rights and female gendered, 119, 121, 127, 131, 137, 140–41, 144–45; race and, 114, 144–45, 246; sex (work) and, 102,

108–10, 115, 149–57 (science
fiction visions of, 157–62); social,
170–76, 179–81, 183–85
Rothblatt, Martine and Bina, 113–14.
See also Terasem
Ruberg, Bonnie, 64–66, 71. *See also*
queer (gaming)

Salamon, Gayle, 197–201, 205–7. *See
also* transgender (studies)
Saudi Arabia: women's rights and, 119,
127, 131, 136, 140–41. *See also*
Sophia (robot citizen)
Savulescu, Julian, 105–7. *See also*
transhumanism
Schneider, Rebecca, 201–2, 205. *See
also* performance (art)
sexuality, 40, 50, 70, 170; gender and,
109–10, 206–8; hetero-, 252; incel
hoarding of and right to, 155–56;
male terror at female, 218; robot and
AI, 7–8, 149–50, 157–62. *See also*
erotechnics; robosexuality
simulation, 9; gamification and, 59,
239–41, 253; science fiction and,
221–22; sex robots and, 152; of
synesthetic environments, rave
culture and cyberspace, 239–41,
253; in Virtual Reality (VR),
248–50
social justice, 70, 195, 219;
gamification and, 64; robots and, 6;
transhumanism and, 102, 112–14
social media, 2–5, 138, 153; aging and,
40, 42, 44, 54–55; gamification and,
63, 67; weak regulation, advertising
discrimination and, 17–24, 227. *See
also* Facebook
Sophia (Hanson Robotics), 119–45; as
citizen of Saudi Arabia, 4, 7, 119,
127, 131, 136, 140–41. *See also*
robots (social)
SOPHIE (electronic musician), 196,
202, 205, 208

supercrip, 4
synesthesia: discussion and definition
of, 237–39, 242; rave culture and,
241–42. *See also* affect (theory); *Rez*;
in (video) gaming (*Rez*), 240–45,
247–54

Tamagotchi, 173. *See also* robots
(social)
Terasem, 113–14, 125. *See also*
transhumanism
Trance Vibrator (*Rez*), 238–40, 243–44,
248–52
transbeman, 113. *See also* Terasem
transgender: embodiment, 199;
feminists, 113, 115; studies,
195–97
transhuman(ism): definition and
discussion of, 99–105, 125,
129; democratic, 110–14;
libertarian, 105–10, 254. *See also*
posthuman(ism)
transparency: Facebook's lack of, 23–
27, 33; mandating, 30, 32, 90
TrueCompanion, 150. *See also* robots
(sex)
Turkle, Sherry, 150, 171, 173, 175,
177–78
Twitter, 44, 63, 135, 143. *See also*
social media

video game(s): moral panic
over violence in, 66; queer
narratology and, 238, 247–48,
253; science fiction and, 236. *See
also* Gamergate; game studies;
gamification; *Rez*
Virtual Reality (VR): Cartesian
cyberspace of, 203; clinical
therapeutic use of, 244; *Rez* (video
game) and, 9, 238, 243–44, 248, 252;
synesthesia and, 251
vulnerability: human, 169–70;
intersectionality and, 170; layers

of, 170–71; transhumanism and, 103

Wagner, Richard, 238. *See also Gesamkunstwerk*
Westworld (television show), 4, 151, 158, 160, 161

xenofeminism: definition and discussion of, 9, 213–15, 217, 219–20; reproductive technologies and, 224, 228, 230–32. *See also* abolition; alienation
xenomorph, 215–18, 220–23, 232. *See also* alien(s)

About the Contributors

Joep Bouma, MA (1993) is a lecturer in New Media & Digital Culture at the University of Amsterdam. Their research has been devoted to an intersection of disciplines around transgender identity and activism within media studies. Closely affiliated with Amsterdam's arts and culture sector, artistic research constitutes a central part of their work. Practices of online (vernacular) creativity by gender minorities that employ digital means for a variety of social and personal purposes constitute their main object of research interest. Their analyses merge the conceptual and the material and critically assess the potentialities of new media technologies from queer and trans perspectives. The primary fields that inform their research are feminist media studies, digital art and culture, new media philosophy, and queer theory. In terms of research ethics, especially as their research concerns marginalized individuals and communities, they take position as an engaged researcher with interdisciplinary qualitative methods.

Jordan Canzonetta is an assistant professor of English Studies at Lewis University in Romeoville, Illinois. She earned her PhD in composition and cultural rhetoric from Syracuse University. Her work has appeared in the *Journal of Writing Assessment* and in international and national edited collections on digital rhetorics and plagiarism detection technologies. Her expertise is squarely located in rhetorical studies related to the digital sphere, AI and automation, technical communication, academic labor, and writing assessment in higher education.

Christopher M. Cox is a lecturer in the Creative Industries Faculty at the Queensland University of Technology (QUT) and a chief investigator at the Digital Media Research Centre (DMRC). His research focuses on automated and autonomous media technologies, with a particular emphasis on industrial ideologies of autonomy shaping the development and cultural inception of automated media. To date, his work has appeared in the scholarly journals *Convergence: The International Journal of Research into New Media Technologies, Journal of Digital Media and Policy, Critical Studies in Television, Transformative Works and Cultures*, and *tripleC: Communication, Capitalism & Critique*, in addition to multiple chapters in edited book volumes.

Scott DeJong is a PhD student in Communication Studies at Concordia University examining serious play, web literacy, and digital cultures. His past work looked at educational game design and online echo chambers. Currently, he is investigating social class within games, the simulation of politics within games, and relationships between age and technology. His work focuses on the role of play in mediating specific discourse and interactions between players and a system or object. He has written some posts around class presentations in games for the *Class and Games* research blog. He also co-authored *Social Justice Games: Building an Escape Room on Elder Abuse through Participatory Action Research* (2020) and took a lead role in designing the game that the article discusses. This piece, alongside his master's thesis, focuses on blending theory with practice in order to discuss serious social issues. Scott is an active member of the Ageing + Communication + Technologies (ACT) project, the TAG (Technoculture, Arts and Games) Lab, Scaling Liveness in Participatory experiences, the Algorithmic Media Observatory, and the mLab.

Julia A. Empey is a doctoral candidate at Wilfrid Laurier University. She completed her Hon. Bachelor of Arts in English and History with a Religious Studies minor and her Master of Arts in Cultural Studies and Critical Theory at McMaster University. Her graduate studies are funded through a Joseph-Armand Bombardier Canada Graduate Scholarship. Julia's doctoral research utilizes posthumanism along with feminist discourses to examine the cultural and social implications of sex robots. Her other interests include eco-criticism, cosmopolitan studies, and political theory.

Jamie Foster Campbell is a PhD candidate at the University of Illinois at Chicago in the Department of Communication. Her work focuses on interpersonal communication and technology use. Specifically, she studies how technology, in all its iterations, is a factor in establishing intimacy. By taking a step back from looking at particular channels of communication people use to develop and maintain their relationships, her research focuses on how the architecture of technology shapes acts of intimacy. Jamie holds a Bachelor of Arts from the University of San Francisco and a Master of Arts from San Francisco State University in communication. She has presented her work at various academic conferences including the International Communication Association, the Association of Internet Researchers, the International Association of Relationship Researchers, the National Communication Association, and the Western States Communication Association.

Maude Gauthier works in the media industry and collaborated on various projects that examined the mediated experiences of adults in later life with the Aging + Communication + Technologies (ACT) project. She received her PhD in communication studies from the Université de Montréal in Canada.

Sebastián Gómez (no pronoun/pronoun indifferent), philosopher and researcher specialized in digital cultures, is a doctoral candidate as part of the *Gamification Lab* at the Institute of Culture and Aesthetics of Digital Media (ICAM), Leuphana University of Lüneburg, Germany. Adopting a posthumanist framework, Sebastián explores the material configurations of gamification in the perpetuation of structures of dominance aiming at the identification of strategies for intersectional justice.

Kristina M. Green is a PhD candidate at the University of Illinois at Chicago in the Department of Communication. Her research examines the sociocultural and ethical consequences of human–machine communication, big data, surveillance capitalism, and AI. Kristina is a qualitative researcher who specializes in case study research design, semi-structured interviews, and walkthrough approaches to smartphone apps and infrastructure. She holds a Bachelor of Science in Film and Television Production and a Master of Arts in Emerging Media Studies from Boston University. Kristina has presented research at academic conferences (National Communication Association, the Association of Internet Researchers, and the Society for Literature, Science, and Art) and international workshops at the University of Milan in Italy, Queensland University of Technology in Australia, and the Oxford Internet Institute in England.

Nikila Lakshmanan is a citizen of the United States. She is a 2016 graduate of Smith College in Northampton, Massachusetts. Nikila received a double major in philosophy and women's studies, and her main research interest is in feminist theory. Nikila is currently a PhD student in the program in Jurisprudence and Social Policy (JSP) at U.C. Berkeley.

Madelaine Ley is a PhD candidate at the Delft University of Technology. Trained in philosophy and science and technology studies, Ley works at the AIforRetail Lab (AIRLab) researching ethics and the future of work with robots. Ley adapts feminist and care ethics theory to the context robots in retail, highlighting the importance of understanding people's embodiment, emotions, and sociality at work. Together with engineers, Ley collaborates to incorporate ethics proactively during the beginning stages of robot design. In addition to her work at AIRLab, Ley also researches the ethics of robot-touch and intimacy with robots.

Chloé Nurik is a PhD/JD candidate at the Annenberg School for Communication and the University of Pennsylvania Law School (where she is a Levy Scholar). She researches freedom of expression protections and limitations, business ethics, self-regulation of media industries, and gender discrimination. Her work has been published in the following journals: *International Journal of Communication*; *Communication, Culture and Critique*; *Information Research*; *Communication and the Public*; *Sexual Health*; and *Electronic Journal of Communication*. Chloé has presented her research at the Oxford Media Policy Summer Institute: Technology & Policy at the Margins (Oxford), International Communication Association Annual Conference (Washington, D.C. and Prague), Internet, Policy & Politics Conference (Oxford), Milton Wolf Seminar on Media and Diplomacy (Vienna), The Changing Facets of Evil and Free Speech Conference (Lisbon), and the Eastern American Studies Association Annual Conference (Harrisburg). Her dissertation analyzes how the self-regulation of social media sites impacts historically marginalized groups and communities. Chloé has an MA in communication and a BA in history (with a gender concentration) from the University of Pennsylvania.

Nathan Rambukkana is an assistant professor in Communication Studies at Wilfrid Laurier University, in Waterloo Canada. His work centers the study of discourse, politics, and identities, and his research addresses topics such as the "alt-right"; hashtag publics; digital, haptic, and robotic intimacies; intimate privilege; and non/monogamy in the public sphere. He is the author of *Fraught Intimacies: Non/Monogamy in the Public Sphere* (UBC Press, 2015) and the editor of *Hashtag Publics: The Power and Politics of Discursive Networks* (Peter Lang, 2015). http://complexsingularities.net

Kim Sawchuk is a professor in the Department of Communication Studies at Concordia University. She is the director of Ageing, Communication, Technologies: Experiencing a Digital World In Later Life (ACT). Her research asks what it means to age in a society where the pressure to become digital is being made into an imperative for participation in public life. Kim's most recent work on aging and media is centered on community-based media practices with older adults and is asking questions about the ways in which Web 3.0 is shaping public knowledge of age and ageing.

tobias c. van Veen is visiting professor in Politics at Acadia University. He holds doctorates in Communication Studies and Philosophy from McGill University. His research addresses philosophy of race, sound, and technology in critical media studies, and he has published widely on Afrofuturism, post-humanism, and electronic dance music cultures (EDMC). Tobias is co-editor of the Afrofuturist Studies & Speculative Cultures series at Lexington Books; co-editor with Reynaldo Anderson of the "Black Lives, Black Politics, Black Futures" special issue of *TOPIA: Canadian Journal of Cultural Studies* (2018); editor of the Afrofuturism special issue of *Dancecult: Journal of Electronic Dance Music Culture* (2013); and co-editor of the special issue Echoes from the Dub Diaspora (2015). He hosts the *Other Planes: Afro/Futurism* podcast on CreativeDisturbance.org and DJs at twitch.tv/pande-mixDJs.

www.ingramcontent.com/pod-product-compliance
Lightning Source LLC
Chambersburg PA
CBHW051955270326
41929CB00015B/2667